Texts in Algorithmics

Volume 6

Algorithms

in

Bioinformatics

Volume 1
Handbook of Exact String Matching Algorithms
Christian Charras and Thierry Lecroq, eds

Volume 2
String Algorithmics
Thierry Lecroq and Costas Iliopoulos, eds

Volume 3
CompBioNets 2004: Algorithms and Computational Methods for Biochemical and Evolutionary Networks
K. S. Guimarães and M.-F. Sagot, eds

Volume 4
Algorithms and Complexity in Durham 2005
H. Broersma, M. Johnson and S. Szeider, eds.

Volume 5
CompBioNets 2005: Algorithms and Computational Methods for Biochemical and Evolutionary Networks
M.-F. Sagot and K. S. Guimarães, eds

Volume 6
Algorithms in Bioinformatics
C. S. Iliopoulos, K. Park and K. Steinhöfel, eds

Texts in Algorithmics Series Editor
Costas S. Iliopoulos   csi@dcs.kcl.ac.uk

Algorithms in Bioinformatics

Edited by
Costas S. Iliopoulos,

Kunsoo Park

and

Kathleen Steinhöfel

© Individual author and King's College 2006. All rights reserved.

ISBN 1-904987-36-2
College Publications
Scientific Director: Dov Gabbay
Managing Director: Jane Spurr
Department of Computer Science
Strand, London WC2R 2LS, UK

Cover design by Richard Fraser, www.avalonarts.co.uk
Printed by Lightning Source, Milton Keynes, UK

All rights reserved. No part of this publication may be reproduced, stored in a retrieval system or transmitted, in any form, or by any means, electronic, mechanical, photocopying, recording or otherwise, without prior permission, in writing, from the publisher.

Texts in Algorithmics
Volume 6

Algorithms
in
Bioinformatics

Volume 1
Handbook of Exact String Matching Algorithms
Christian Charras and Thierry Lecroq, eds

Volume 2
String Algorithmics
Thierry Lecroq and Costas Iliopoulos, eds

Volume 3
CompBioNets 2004: Algorithms and Computational Methods for Biochemical and Evolutionary Networks
K. S. Guimarães and M.-F. Sagot, eds

Volume 4
Algorithms and Complexity in Durham 2005
H. Broersma, M. Johnson and S. Szeider, eds.

Volume 5
CompBioNets 2005: Algorithms and Computational Methods for Biochemical and Evolutionary Networks
M.-F. Sagot and K. S. Guimarães, eds

Volume 6
Algorithms in Bioinformatics
C. S. Iliopoulos, K. Park and K. Steinhöfel, eds

Texts in Algorithmics Series Editor
Costas S. Iliopoulos   csi@dcs.kcl.ac.uk

Algorithms in Bioinformatics

Edited by
Costas S. Iliopoulos,

Kunsoo Park

and

Kathleen Steinhöfel

© Individual author and King's College 2006. All rights reserved.

ISBN 1-904987-36-2
College Publications
Scientific Director: Dov Gabbay
Managing Director: Jane Spurr
Department of Computer Science
Strand, London WC2R 2LS, UK

Cover design by Richard Fraser, www.avalonarts.co.uk
Printed by Lightning Source, Milton Keynes, UK

All rights reserved. No part of this publication may be reproduced, stored in a retrieval system or transmitted, in any form, or by any means, electronic, mechanical, photocopying, recording or otherwise, without prior permission, in writing, from the publisher.

Preface

The present volume is dedicated to aspects of algorithmic work in bioinformatics and computational biology with an emphasis on string algorithms that play a central role in the analysis of biological sequences. The papers included are a selection of articles corresponding to talks given at one of two meetings sponsored by The Royal Society, the UK's national academy of science, under grant no.: JEB/KOREAN Networks/16715. The grant supported two workshops organised by researches from the Seoul National University (Korea) and King's College London (UK). The first workshop was held in Seoul, Korea, in July 2004 and the second meeting took place in London, UK, in February 2005 as part of the annual London Stringology Days.

The meetings have provided a forum for the presentation and discussion of on-going research and current problems in the field. Especially topics like efficient data structures for the analysis and the storage of biological sequences, the design of efficient algorithms for the detection of various types of repeats in biological sequences, classification of gene sequences, and comparing genomes using repeat detection algorithms were in the centre of the workshop.

The five contributions included in this volume are as follows:

- Nadia Pisanti, Henry Soldano, Mathilde Capentier, and Joel Pothier, *Implicit and Explicit Representation of Approximated Motifs*.

- Tetsuo Asano, Patricia Evans, Ryuhei Uehara, and Gabriel Valiente, *Site Consistency in Phylogenetic Networks with Recombination*.

- Georgios Lappas, Kathleen Steinhöfel, and Andreas A. Albrecht, *Classification of Splice-junction Gene Sequences by a Special Type of Threshold Circuits*.

- Joong Chae Na, Kangho Roh, Alberto Apostolico, and Kunsoo Park, *Sequence Alignment with Quality Scores*.

- Gregory Kucherov, Laurent Noé, and Mikhail Roytberg, *A Unifying Framework for Seed Sensitivity and its Application to Subset Seeds*.

We wish to thank all who supported these meetings, all authors who submitted papers, all members who were in involved in their organisation, and The Royal Society for financial support.

August 2005
Costas S. Iliopoulos
Kunsoo Park
Kathleen Steihöfel

CONTENTS

NADIA PISANTI, HENRY SOLDANO, MATHILDE CAPENTIER, AND JOEL POTHIER
Implicit and Explicit Representation of Approximated Motifs   1

TETSUO ASANO, PATRICIA EVANS, RYUHEI UEHARA, AND GABRIEL VALIENTE
Site Consistency in Phylogenetic Networks with Recombination   15

GEORGIOS LAPPAS, KATHLEEN STEINHÖFEL, AND ANDREAS A. ALBRECHT
Classification of Splice-junction Gene Sequences by a Special Type of Threshold Circuits   27

JOONG CHAE NA, KANGHO ROH, ALBERTO APOSTOLICO, AND KUN-SOO PARK
Classification of Splice-junction Gene Sequences by a Special Type of Threshold Circuits   45

GREGORY KUCHEROV, LAURENT NOÉ, AND MIKHAIL ROYTBERG
A Unifying Framework for Seed Sensitivity and its Application to Subset Seeds   63

Implicit and Explicit Representation of Approximated Motifs

NADIA PISANTI[1], HENRY SOLDANO, MATHILDE CAPENTIER, AND JOEL POTHIER

ABSTRACT. Detecting repeated 3D protein substructures has become a new crucial frontier in motifs inference. In [7] we have suggested a possible solution to this problem by means of a new framework in which the repeated pattern is required to be conserved also in terms of relations between its position pairs. In our application these relations are the distances between α-carbons of amino acids in 3D proteins structures, thus leading to a *structural consensus* as well. In this paper we motivate some complexity issues claimed (and assumed, but not proved) in [7] concerning inclusion tests between occurrences of repeated motifs. These inclusion tests are performed during the motifs inference in *KMRoverlapR* (presented in [7]), but also within other motifs inference tools such as *KMRC* ([9]). These involve alternative representations of motifs, for which we also prove here some interesting properties concerning pattern matching issues. We conclude this contribution with a few tests on cytochrome P450 protein structures.

1 Introduction

Finding repeated subsequences and substructures in biological (resp. sequential and structural) data is having growing importance for various different applications in molecular biology. Among them we can mention the detection of trasncription factors binding sites as repeated gapped motifs in the upstream regions preceeding genes, or the prediction of RNA secondary structures as complementary reversed repeated subsequences, the detection of common fragments of genomic sequences as a starting point of measuring genomic distances, etc. In this paper we focus on yet another biological application, that is the detection of common substructures in 3D proteins. In [7] we have designed an algorithm for the inference of repeated motifs under the new framework of *relational* motifs which results particularly suitable for this purpose.

Motifs inference in biological applications requires a certain degree of approximation in establishing whether a biological object is basically the same as another

---
[1] Supported by the ACI IMPBio *Evolrep* project of the French Ministry of Research.

one. For this reason, the possibly huge size of solutions in the search space makes the algorithmical solution tricky. It is very difficult to find the right balance between the sensitivity of a motif inference tool and its efficiency when an exhaustive algoritmical approach is suited. Most of the difficulty comes from the unavoidable noise of biological data which causes an explosion of intermediate candidates (typically, shorter motifs to be extended or composed to make longer ones). Hence, it is very important that the inference tool offers a way to refine the query in order to minimize this noise. For this purpose, we have designed *KMRoverlapR* that detects repeated motifs that are approximated because they are patterns defined on a input degenerate alphabet, and they are also required to be conserved in terms of relations between each pair of positions of the consensus sequence. In our application to 3D proteins, the input sequences are amino acid sequences enriched with the information, per each pair of positions that are at most at a distance of k letters, of the distance between the corresponding α-carbons in the 3D structure. Moreover, the amino acids are grouped into possibly overlapping subgroups that somehow represent similar physical and chemical characteristics. Finally, also for the distance between the α-carbons, it is given a set of possibly overlapping ranges of values. A relational k-pattern is a k-long sequence of the groups of amino acids among those given above, with the ranges of its $k(k-1)/2$ distances between the α-carbons, in the 3D structure, of each pair of distinct positions. A pattern *occurs* in the input sequence if the latter contains a k-long fragment where in all positions the amino acid belongs to the corresponding group of the pattern, and each pair of them is at a distance in the 3D structures that fits the ranges of the distances required by the pattern's relations. Given a quorum q, the goal is to detect all (relational) k-patterns that occur at least q times, and that we will name (relational) k-motifs.

In [7] we have introduced a linear time algorithm for the inference of repeated relational k-motifs. Thanks to some properties proved in [7], the algorithm guarantees a complete and correct inference avoiding to have to list all candidates in intermediate steps. This is achieved by means of an implicit representation that only uses the *extent* of a motif (*i.e.* the complete set of its occurrences), and thanks to some sufficient conditions that allow to keep only distinct extents all along the computation. In this paper we address some issues concerning this implicit representation and its possible alternatives for some specific purposes. In particular, we will discuss properties of a couple of explicit representation for possible post-processing of the inferred motifs, and we will motivate some complexity issues claimed (but not proved) in [7] that involve the representation of the motifs.

2 Previous Work: KMRC and KMRoverlapR

In [7] we have introduced a tool for inferring approximatively repeated relational motifs. The framework of relational motifs is very powerful in that it may allow refined queries thus leading to sensible and, at the same time, efficient inference of repeated substructures. In [7] we have given motivations for choosing a *KMR*-like approach ([6]) when relations are taken into account [2] which, in turn, leads to the choice of a degenerate alphabet to express the approximation. The degenerate alphabet to be used to describe the motif is explicitly given as an input parameter (a second degenerate alphabet is possibly also given for the relations in *KMRoverlapR*) under the form of a cover G over the alphabet Σ of the input sequence. Each element of this cover is a subset of Σ and we will denote these elements as *groups*. We denote with *degeneracy g* the maximum number of distinct groups to which a letter belongs to. A motif is thus seen as a sequence of groups $C_1 \ldots C_k$ that occurs in the input sequence at position p whenever there is a sequence of letters $\sigma_1 \ldots \sigma_k$ such that $\sigma_j \in C_j \in G$ for $1 \leq j \leq k$. The inference algorithm of *KMRoverlapR* ([7]) is a suitable extension of the *KMRC* one ([9]). In [7] we have addressed some issues that raised specifically for relational motifs in *KMRoverlapR*. Nevertheless, many properties concerning the compact representation of the motifs by means of their extents, the filtering of maximal motifs, as well as complexity issues, are actually shared by the two tools. Among these, there are the issues discussed in this paper. For this reason, and since they can be straightforwardly extended to the case of relations, in what follows we will refer to motifs without relations in order to simplify the notation. We will denote with k-*motif* a motif of length k.

A common feature of *KMRC* and *KMRoverlapR* is the restriction to *maximal* motifs. A maximal k-motif is a motif whose complete list of occurrences, that we will name *extent*, is not properly included into the extent of another k-motif. It is a *duplication* if it is equal. In [9] an upper bound of the total size of the extents of k-motifs has been proved, and in [7] its natural extension to relational motifs is shown. This bound is theoretically the same whether or not we restrict to maximal and non duplicated motifs only. Nevertheless, in practice we observed noticeable ratios between their number (see [7] for details). This, and the fact that maximal motifs of a fixed length suffices to infer all distinct maximal motifs of greater length, motivates the elimination, at each intermediate step, of all non maximal (or duplicated) extents. This requires an exhaustive inclusion test between all pairs of candidate motifs, which becomes actually the bottleneck of the computation.

Omitting variants and specific features due to the introduction of relations, the inference algorithm we refer to can be summarized in the following steps where we assume that we seek maximal k-motifs, that are k-long words of the alphabet

---

[2]That is, an *in width* inference of motifs in the sense that all motifs of length ℓ are inferred before any motif of length $> \ell$. The complementary choice is an *in depth* approach like [8] where each single candidate is extended as long as it satisfies the requirements. In general, when the motif is represented as a consensus pattern, the *in depth* results a better choice ([4]), but in [7] we have shown that with relations the things change.

of the groups, that occur at least q times in an input sequence s of length n.

1. Compute extents of each group, *i.e.* compute extents of $(\ell = 1)$-motifs.
2. **while $\ell < k$ do**
   (a) Compute extents of $(\ell + d)$-motifs from those of ℓ-motifs; $\ell := \ell + d$;
   (b) Eliminate extents containing less than $< q$ occurrences.
   (c) Eliminate extents that are included into others.
3. Output all left extents (that is, all maximal k-motifs).

In other words, it is possible to perform the inference keeping only the extents of the motifs, that are ordered subsets of $\{1, \ldots, n\}$. On these the most involving operation we do is the $2(c)$ above consisting in the detection of extents that are equal to or included into others. We will denote step $2(c)$ as *Inclusion Test*.

3 Representation of Maximal Motifs

3.1 Implicit Representation with Occurrences Lists

The set of distinct patterns of length k can obviously be as big as the set of different k-long words on the alphabet G of the groups, which has size $|G|^k$. For example, in the simple (although improbable) sequence $\overline{\sigma}^n$, if $\overline{\sigma} \in \Sigma$ occurs in all groups of G, then we have that every string in G^k is a k-motif for $1 \leq k \leq n - 1$ (for any quorum $1 \leq q \leq n - k + 1$). Hence, the upper bound mentioned above happens to be tight and therefore an explicit representation of all motifs of a given length is unfeasible. On the other hand, the exponential number of motifs shown above can be represented by an unique extent $X = \{1, 2, \ldots, n - k + 1\}$ and a length k, (that is in linear space). This is, intuitively, the motivation of why the algorithm of *KMRC* and *KMRoverlapR* actually deal with extents only. In fact, the above mentioned motifs of the sequence $\overline{\sigma}^n$ can all be represented by an unique extent because they are all maximal duplications of each other. We observed (see [7]) a ratio in $O(10^3)$ between the number of maximal motifs and the nonmaximal ones, when the latter are $O(10^5)$. Therefore, in practice the gain is noticeable.

3.2 Explicit Representation of Motifs

The implicit representation described in Section 3.1 allows a sensible speed up in the inference phase, and in particular it avoids an explosion of generated candidates. Nevertheless, for the purpose of describing the output, a more explicit representation would be more suitable, in order to *visualize* the actual motifs once their inference is performed. Moreover, as we will see in Section 4, also some complexity issues motivate a switch to an explicit representation already during the inference phase.

An extent X of a motif actually represents the following set of motifs.

$$M(X) = \{C_1 C_2 \cdots C_k \mid s[p+j-1] \in C_j,\ \forall\, 1 \le j \le k,\ \forall\, p \in X\}.$$

Moreover, we will denote with $M_j(X)$ the set of groups that are at position j in $M(X)$, that is $M_j(X) = \{C_j \mid C_1 \cdots C_j \cdots C_k \in M(X)\}$ for any $1 \le j \le k$.

EXAMPLE 1.1. In the simple sequence $s = abbbc$ with the cover $G = \{C_1 = \{a,b,z\}, C_2 = \{b,c,z\}, C_3 = \{x\}\}$, we have that the extent $X = \{1,2,3\}$ of a 3-motif represents the set of motifs $M(X) = \{C_1 C_1 C_2, C_1 C_2 C_2\}$ and that $M_2(X) = \{C_1, C_2\}$.

The motifs set $M(X)$ is itself an explicit representation, but it can results too redundant. There are more compact ways — still more explicit than the extents — to represent $M(X)$. We report here below two possibilities.

1. **G-representation.** A k-long sequence of intersections of groups of the cover G. For each position $1 \le j \le k$ we have

$$G_X[j] = \cap_{C_j \in M_j(X)}\, C_j.$$

2. **Σ-representation.** A k-long sequence of subsets of Σ, listing for each position $1 \le j \le k$ the set $\Sigma_X[j]$ of letters occurring at position j in the occurrences. Formally:

$$\Sigma_X[j] = \cup_{p \in X}\, s[p+j-1].$$

EXAMPLE 1.2. Let us consider again the input text $s = abbbc$, the cover $C_1 = \{a,b,z\}, C_2 = \{b,c,z\}, C_3 = \{x\}$, and the extent $X = \{1,2,3\}$ representing $M(X) = \{C_1 C_1 C_2, C_1 C_2 C_2\}$. Its G-representation is $G_X = C_1(C_1 \cap C_2)C_2 = \{a,b,z\}(\{a,b,z\} \cap \{b,c,z\})\{b,c,z\} = \{a,b,z\}\{b,z\}\{b,c,z\}$, and the Σ-representation is $\Sigma_X = \{a,b\}\{b\}\{b,c\}$.

There are other possible representations, among which we can mention position specific scoring matrices (that is, a $|\Sigma| \times k$ table where for each $\sigma \in \Sigma$ and for each $1 \le j \le k$ we report the number of times the letter σ occurs at position j in an occurrence of the motifs set) as well as a variant containing the same information in terms of groups. The information of the distribution of the letters/groups could also be added in the G- and Σ- representations by using multisets rather than simple sets. These could be suitable for applications in which the statistics of the distributions of the letters is useful, and it can even result efficient for small size alphabets. These two conditions may hold for consensus sequences in DNA or RNA sequences. Nevertheless, in this paper we will concentrate on the two G- and Σ- representation formalized above because, as we will see in Section 4, they raise interesting complexity results for some crucial steps in the inference of motifs performed by *KMRoverlapR*.

The G- and Σ- representations are somehow related in that they eventually both display the motif as a sequence of subsets of Σ. Notice that, even if the Σ-representation is somehow independent from G, this latter has driven the inference and hence the resulting output. As a consequence, there are a few relations among the G-representation, the Σ-representation, and the cover G.

LEMMA 1.3. *For all* $1 \leq j \leq k$, $\Sigma_X[j]$ *is a subset of at least one group of* G.

Proof. By definition, X is the extent of at least one motif $C_1 \cdots C_J \cdots C_k$, and hence $s[p+j-1] \in C_j \ \forall p \in X$ and thus $\Sigma_X[j] \subseteq C_j$. ∎

Actually, for the very same reason we have that $\Sigma_X[j] \subseteq C_j$ for each distinct C_j whose intersection is $G_X[j]$, which leads to the following result that is a direct consequence of Lemma 1.3.

PROPOSITION 1.4. *For all* $1 \leq j \leq k$, $\Sigma_X[j] \subseteq G_X[j]$.

Depending on the application, the output may require that also the cover G is given in order to reconstruct lost information. We will discuss in Section 3.3 some pattern matching issues resulting from some loss of information in the different representations.

3.3 Searching Occurrences in a New Text

One of the possible need of an explicit representation of inferred motifs is the post-processing of such data. For example, in biological applications, the patterns resulting from motifs inference are often object of successive queries in pattern matching in order to check their occurrences in a new text. It is clear that if we want to search occurrences of an inferred motif into a new text, we need to process the extent and write an explicit representation of *what* we want to search. In this section we address some issues concerning such queries considering the distinct possible representations we suggested in this paper.

EXAMPLE 1.5. Let us consider the same cover $G = \{C_1 = \{a, b, z\}, C_2 = \{b, c, z\},$
$C_3 = \{x\}\}$, $k = 3$, and the input string s as in the previous example. For the extent $\{1, 2, 3\}$, the G-representation is $G_X = \{a, b, z\}\{b, z\}\{b, c, z\}$, and the Σ-representation is $\Sigma_X = \{a, b\}\{b\}\{b, c\}$. Let us now consider the new text $s' = a\underline{zc}xxx\underline{aab}xxx\underline{abc}$ and, in particular, the underlined substrings of length 3 occurring, respectively, at positions 1, 7, and 13. If we search the G-represented pattern

$\{a,b,z\}\{b,z\}\{b,c,z\}$, we would only find the occurrences 1 and 13. Moreover, searching the Σ-represented $\{a,b\}\{b\}\{b,c\}$ we get position 13 only. Nevertheless, notice that an *ex-novo* inference of maximal 3-motifs occurring at least three times in s' and written in the alphabet of the cover G would result in the extent $\{1,7,13\}$ representing the k-motif $\{C_1\}\{C_1\}\{C_2\}$.

The example has shown that the two representations behave in general differently in possible post-inference text search of a motif. Moreover, they both miss occurrences with respect to a possible ex-novo inference with the same parameters. However, depending from the application, it can be that what one actually wants to find is not the complete set of occurrences as if the motif were inferred from scratch, but rather the possible position where a specific instance of it occurs. For example, assume that we have inferred an over represented fragment in a set of 3D protein structures. Assume that for the spatial distance we have been using an alphabet that groups a range of possible distances in the interval $[d_{min}, d_{max}]$, but that we have detected a frequent substructure having always basically the same distance $\overline{d} \in [d_{min}, d_{max}]$ between two specific positions. It is reasonable to think that after such inference one wants to search this specific observed pattern. In this sense explicit representations with loss of information such as the two above can still result as valid.

3.4 Complexity Issues of Explicit Representation

In this section we analyse time and space complexity of computing and storing the two different explicit representations.

Computing the G-representation requires an exhaustive search in all positions of all occurrences of the motifs X represents. And per each one of them the degeneracy of G has to be taken into account as well. We assume that we have a vector V containing, for each $1 \leq i \leq n$, the set $V[i]$ of groups occurring at position i of the input sequence[3]. By definition of V and $G_X[j]$, we have that, for all $p \in X$, if $s[p+j-1] \in C_j$, then $C_j \in V[p+j-1]$. Nevertheless, for different $p \in X$ there are obviously different sets $V[p+j-1]$, each one being in general a superset of $M_j(X)$ and thus of $G_X[j]$. We have the following useful result.

PROPOSITION 1.6. $G_X[j] = \cap_{p \in X} V[p+j-1]$.

Proof. We have that $G_X[j] \subseteq \cap_{p \in X} V[p+j-1]$ as a direct consequence of the fact that, $\forall\, p \in X$, if $s[p+j-1] \in C_j$ then $C_j \in V[p+j-1]$. For the opposite $(\cap_{p \in X} V[p+j-1] \subseteq G_X[j])$ let us fix j and consider any $C_j \in G$ such that $C_j \in \cap_{p \in X} V[p+j-1]$. By definition of V this means that $s[p+j-1] \in C_j$

---
[3]This data structure is actually created both in *KMRC* and in *KMRoverlapR* and kept during the inference phase.

$\forall p \in X$ and thus that $C_j \in M_j(X)$. Hence, $G_X[j]$ contains the intersection of all such C_j's proving the thesis. ∎

As a consequence of Proposition 1.6, $G_X[j]$ can be computed as the intersection of $|X|$ lists of groups, each one ($V[p+j-1]$) containing at most g elements. Such lists are ordered and $\cap$ is associative, and thus it suffices to perform a linear visit to the lists to compute the intersection. Hence, given that $\sum_X |X| \leq ng^k$, computing the G-representations of all the extents X of maximal k-motifs takes $\sum_X |X| \cdot k \in O(ng^k k)$ time in the worst case. The space complexity is also in $O(ng^k k)$.

The Σ-representation of all maximal k-motifs can be computed, for all extents X, and for all positions $1 \leq j \leq k$, by doing the union of $s[p+j-1] \, \forall \, p \in X$, which can result in at most $|\Sigma|$ elements, then taking $\sum_X |X| \cdot |\Sigma| \in O(ng^k |\Sigma|)$ time and space.

4 Inclusion Test with Explicit Representation

Both in *KMRC* and in *KMRoverlapR* the bottleneck is the motifs inference in the elimination of non-maximal motifs. This requires an exhaustive search in extents included into others, and the inefficiency is caused by the fact that all extents must be pairwise tested for a possible inclusion, each one of them can contain as many as n elements. This drawback could be avoided with an explicit representation of motifs because the comparison would be performed between objects of size at most n. In this section we show necessary and sufficient conditions on the explicit representations that correspond to inclusion among extents. Let us start with observing the following fact which is a direct consequence of the fact that extents inferred by *KMRC* and *KMRoverlapR* are the complete set of occurrences of one or more maximal motifs.

FACT 1.7. *Any maximal extent X extracted by KMRC or KMRoverlapR from a sequence s has the property that there exists no $\overline{p}$ such that $\overline{p} \notin X$ and $s[\overline{p}+j-1] \in \cup_{p \in X} s[p+j-1] \, \forall \, 1 \leq j \leq k$.*

We give now a necessary and sufficient condition to detect nonmaximal or duplicated motifs within the explicit representation. In what follows, we will say that $\Sigma_{X'} \subseteq \Sigma_X$ if $\Sigma_{X'}[j] \subseteq \Sigma_X[j] \, \forall \, 1 \leq j \leq k$, and similarly that $G_X \subseteq G_{X'}$ if $G_X[j] \subseteq G_{X'}[j] \, \forall \, 1 \leq j \leq k$. Let us start observing that $X' \subseteq X \iff M(X) \subseteq M(X')$ because adding positions p where a motif is required to occur can only decrease the set of motifs satisfying the condition.

LEMMA 1.8. *Let X, X' be extents of k-motifs. We have that*

$$X' \subseteq X \iff \Sigma_{X'} \subseteq \Sigma_X.$$

Proof. If $X' \subseteq X$, then we have that $\Sigma_{X'}[j] = \cup_{p \in X'} s[p + j - 1] \subseteq \cup_{p \in X} s[p + j - 1] = \Sigma_X[j] \ \forall \ 1 \le j \le k$.
If $\Sigma_{X'} \subseteq \Sigma_X$ then we have that $\cup_{p \in X'} s[p + j - 1] \subseteq \cup_{p \in X} s[p + j - 1]$ for all $1 \le j \le k$. Hence it must be that $M(X) \subseteq M(X')$ and thus that $X' \subseteq X$. ∎

Actually, a slightly stronger result (although not useful for the purpose of this section) than Lemma 1.8 holds. Namely, we have that $\Sigma_{X'} \subsetneq \Sigma_X \iff X' \subsetneq X$ because if $\exists \ j'$ and $\tilde{\sigma} \in \Sigma$ such that $\tilde{\sigma} \in (\Sigma_X[j'] \setminus \Sigma_{X'}[j'])$, then we have a position $\tilde{p} \in X$ such that $s[\tilde{p} + j - 1] = \tilde{\sigma} \notin \cup_{p \in X'} s[p + j' - 1]$, which implies that $\tilde{p} \notin X'$ and thus that $\tilde{p} \in (X \setminus X')$.

In terms of G-representation, we have a similar result.

LEMMA 1.9. *Let X, X' be extents of k-motifs. We have that*

$$X' \subseteq X \iff G_{X'} \subseteq G_X.$$

Proof. If $X' \subseteq X$, then we have that $M(X) \subseteq M(X')$ and thus that $\forall \ 1 \le j \le k$ $G_{X'}[j] \subseteq G_X[j]$ because, in general, in $M_j(X')$ there are at least as many groups to intersect as in $M_j(X)$, and further intersections can only decrease the final set. These implications can be easily reversed in case of equality all over.
We still need to prove that $G_{X'} \subsetneq G_X \Rightarrow X' \subseteq X$. The hypothesis implies that there exists one or more j' such that $G_{X'}[j'] \subsetneq G_X[j']$ (and in other positions $G_{X'} = G_X$). This means that $\cap_{C \in M_{j'}(X')} C \subsetneq \cap_{C \in M_{j'}(X)} C$, and hence that $M_{j'}(X) \subsetneq M_{j'}(X')$ and in general $M(X) \subseteq M(X')$, which implies $X' \subseteq X$. ∎

EXAMPLE 1.10. Let us consider the string $xbxcxaxbxc$ and the cover $C_1 = \{a, b\}, C_2 = \{b, c\}, C_3 = \{x\}$. We have that $X = \{2, 6, 8\}$ and $X' = \{2, 8\}$ are such that $X' \subseteq X$ and in fact $M(X) = \{C_1 C_3 C_2\}$ and $M(X') = \{C_1 C_3 C_2, C_2 C_3 C_2\}$. We have that $\Sigma_{X'} = \{b\}\{x\}\{c\} \subset \Sigma_X = \{a, b\}\{x\}\{b, c\}$ and $G_{X'} = \{C_1 \cap C_2\}\{C_3\}\{C_2\} \subset G_{X'} = \{C_1\}\{C_3\}\{C_2\}$.

Lemma 1.9 and 1.8 allow to conceive a different way to perform inclusion tests in order to detect and discard duplicated and nonmaximal k-motifs. Besides the use of explicit representation, the idea of this fast inclusion test is to compare only extents that share a position. This is done by ranging over all positions of the input strings and for each position i we only compare pairs of k-motifs that both occur at position i.

PROPOSITION 1.11. *Detecting nonmaximal and duplicated extents can be done in $O(ng^{2k}k\ell)$ time, where $\ell = min\{g, |\Sigma|\}$.*

Proof. In order to detect pairs of extents that are equal or included one into the other, the necessary and sufficient conditions of Lemma 1.9 and 1.8 allow to compare explicit representations.

Using the G-representation requires to check, for each position in the input sequence (there are n of them), per each pair of motifs both occurring in that position (there are g^{2k} of them), and per each one of their k positions, whether the two ordered lists of at most g elements are included one into the other. The resulting time complexity is in $O(ng^{2k}kg) = O(ng^{2k+1}k)$.

Similarly, with the G-representation we should do, per each pair of motifs and per each one of their k positions, an inclusion test between two ordered sets of size at most $|\Sigma|$. As a result, we need $O(ng^{2k}k|\Sigma|)$ time. ■

Of course, if at each intermediate step of the inference the inclusion step is performed on an explicit representation of motifs, then this latter has to be computed and the cost of this computation must then be taken into account. Nevertheless, this still results into a worst case complexity in $O(ng^{2k}k|\Sigma|^2)$ which is an improvement over the time cost in $O(n^2g^{2k})$ of the inclusion tests performed over the extents because $k, g, |\Sigma| << n$. Indeed, the cost of inclusion test, and hence of the whole inference, becomes linear in the size of the input sequence, thus eliminating the drawback of the quadratic time complexity with respect to the size n of the input sequence.

5 Applications to 3D Protein Structures

As mentioned in the section 1, relational motifs can represent structural motifs in 3D proteins structures. For this purpose the relation between two points x_p and x_q is obtained by discretizing the euclidian distance $d(x_q, x_p)$. Relational motifs represent then geometrical motifs in a multidimensional space, i.e. motifs whose occurrences are insensitive to translations and rotations. Such internal distances between atoms were first used to search structural motifs in the general context where all the atoms of the protein are considered here in [1] and, using a tolerance as here, in [2]. When only considering the α-Carbons in the 3D structure of the protein we obtain a sequence of points in a 3D space.

Here we consider that a prior discretization of the distances has been performed and that relations are denoted as positive integers. We consider a set of relational groups $\{R_j = \{j, ..., j + \delta\}\}$ where δ represents a tolerance level: two discretized distances $d(x_p, x_q)$ and $d(x_{p'}, x_{q'})$ belong to the same group whenever $|d(x_p, x_q) - d(x_{p'}, x_{q'})| \leq \delta$. Note that as a consequence we have that the degeneracy of the relations's alphabet is $\delta + 1$.

Hereunder we give an example of the results obtained when searching a structural pattern repeated in the backbone of several proteins. We chose to study structures of the cytochrome P450 multigenic superfamily (CYP, P450). They

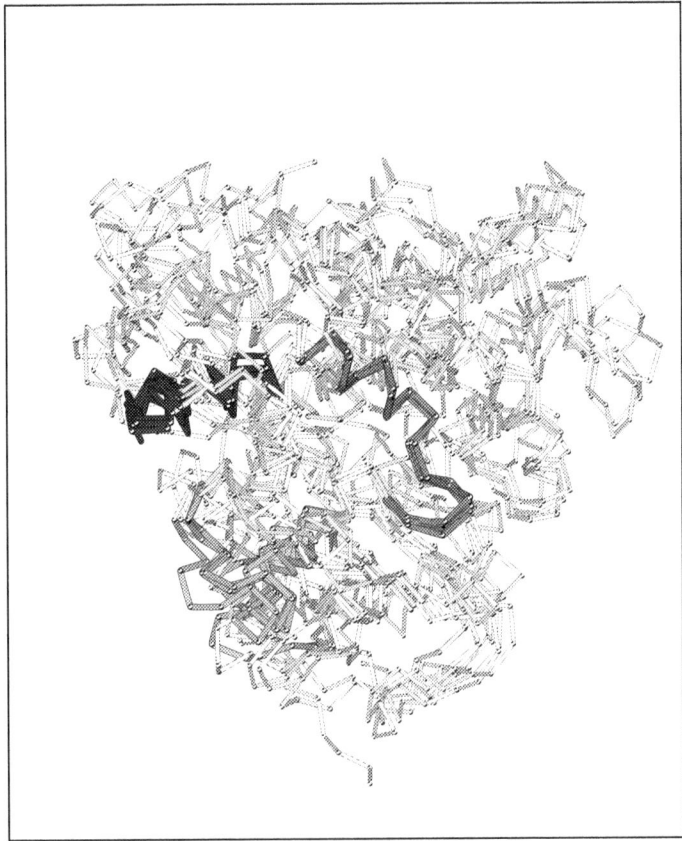

Figure 1. Structural relational motifs of length 18 for five cytochromes P450; their PDB codes are: 4CP4 ,1ROM, 1CPT, 2HPD chain B and 3CPP. This is the maximum length reached using 0.5Å-long intervals, a tolerance level $\delta = 2$ (shorter motifs are not shown). Protein backbones are in grey and motifs are colored and thicker; only α carbons are represented (white small balls) and we trace pseudo-bonds between them (scale 3.8Å between two consecutive C_α). As these locally matching substructures could have slid in one structure with respect to another structure, they may not be aligned all at once. Here structures are aligned according to the green motif.

are heme-thiolate proteins involved in many oxidations of hydrophobic substrates. The substrates are steroid hormones, extracellular fatty acids signaling molecules

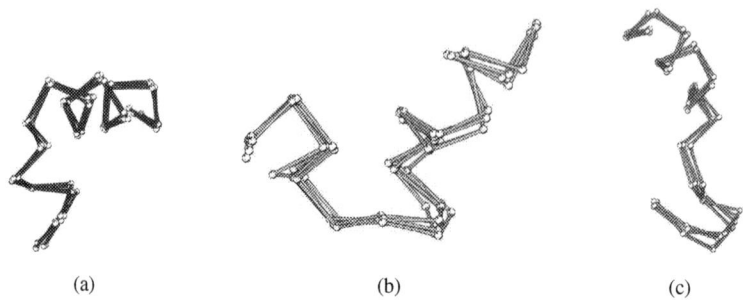

Figure 2. Occurences of the three structural relational motifs of 18 residues. All of them are also found in [5]. They are composed of a coil and a part of α helix. The first motif is the beginning of a very long one going through P450. The last one includes a well conserved Cysteine which bind the heme (7th residue of the motif).

and vitamins but also exogenous substrates as drugs or environmental pollutants (see [3] for an historical review). These P450s can be found in many living beings: bacteria, yeast, fungi, plants, insects, fishes and mammals. They have been widely investigated notably because of their role in drugs degradation. Their amino-acids primary sequences are dissimilar in spite of their structural similarities.

We chose five cytochrome P450 structures: four from bacteria (PDB codes 4CP4, 1CPT, 2HPD chain B and 3CPP) and one from fungi (PDB code 1ROM). Note that here the algorithm searches for patterns that occur at least q times in a set of m protein structures. Here $m = q = 5$. The distances between α-Carbon are discretized using 0.5Å-long intervals, with a tolerance level $\delta = 2$ (so that the degeneracy is 3). The average length of the sequences considered is about 400 residues. There are 10 relational groups (amongst 40) representing distances actually appearing in the sequences. We represent hereunder in Figure 1 the occurrences of the longest structural motifs (k=18) found on the 5 proteins. Such motifs were previously identified on 3 of these proteins [5]. We also show in the table 1 a partial Σ-representation (because only the relations are given) of a motif. As the motif is relational, the Σ-representation is a set of $k(k-1)/2$ constraints, each one expressed as a distance interval δ_{ij}: in order to find an occurrence of this motif at position p in the backbone of a protein, for any pair of positions (i, j) in the motif, the distance between the $p + i^{th}$ and $p + j^{th}$ $\alpha - Carbons$ of the protein has to belong to δ_{ij}.

Table 1. Internal distance intervals (Å) computed for the six first residues of the third motif (green one on figures 1 and 1.2(c)). The distance between two consecutive α-carbons is always nearby 3.8Å, therefore matrix diagonal values are from 3.7Å to 3.9Å. On figure 1.2(c) the first residue is at the bottom and, due to the coil, α-Carbons 1 and 2 are closer to the α-Carbon 6 than to the α-Carbons 3, 4 and 5. As distances are discretized using 0.5Å-long intervals and the tolerance level is $\delta = 2$, distance differences are always less than 1.5Å.

Residue index	2	3	4	5	6	7
1	[3.8 − 3.8]	[6.9 − 7.2]	[8.7 − 9.1]	[6.7 − 7.4]	[4.4 − 5.0]	[6.2 − 6.5]
2		[3.7 − 3.8]	[6.6 − 7.0]	[5.9 − 7.0]	[4.4 − 5.2]	[7.8 − 8.2]
3			[3.8 − 3.8]	[5.4 − 5.7]	[5.3 − 6.0]	[9.1 − 9.7]
4				[3.8 − 3.9]	[5.5 − 5.8]	[8.6 − 9.1]
5					[3.8 − 3.9]	[6.1 − 6.7]
6						[3.8 − 3.8]

6 Conclusion

The method discussed here has been applied to the problem of matching substructures in several protein structures, and the results have been satisfying. As a test case, the matching substructures problem allowed direct (visual) inspection of the fitness of the algorithm, as similar relational motifs are 3D-matching substructures. In this case, the groups of relations - to be considered in the building of relational motifs - are computed as ranges of distances. More generally, relational motifs would be of interest in the biological sequences field, as not only the letters (residues) of the sequences are important, but also relations between some pairs of residues composing a relevant biological motif. And these relations can be not computed but assessed. For a simple example, one can cite amphiphatic helices comparison: in such an helix, a residue has more or less the same hydrophobic index than its neighbours have, and has an hydrophobic index opposite to the residues located at position +4 or -4. The simple relations used here would be "to have similar hydrophobic index" or "to have opposite hydrophobic index". Indeed, relations to be used in the biological sequences field can be much complex than those used in this example.

BIBLIOGRAPHY

[1] C. W. Crandell and D. H. Smith. Computer-assisted examination of compounds for common three-dimensional substructures. *Journal of Chemical Information and Computer Sciences*, 23(4):186–197, 1983.

[2] V. Escalier, J. Pothier, H. Soldano, and A. Viari. Pairwide and multiple identification of three-dimensional common substructures in proteins. *Journal of Computational Biology*, 5(1):41–56, 1998.

[3] R. W. Estabrook. A passion for p450s (rememberances of the early history of research on cytochrome p450). *Drug Metab Dispos*, 31(12):1461–73, 2003. 0090-9556 Historical Article Journal Article Review Review, Tutorial.

[4] M. Tompa et al. Assessing computational tools for the discovery of transcription factor binding sites. *Nature Biotechnology*, 23(1):137–144, 2005.

[5] P. Jean, J. Pothier, P. M. Dansette, D. Mansuy, and A. Viari. Automated multiple analysis of protein structures: application to homology modeling of cytochromes p450. *Proteins*, 28(3):388–404., 1997.

[6] R. Karp, R. Miller, and A. Rosenberg. Rapid identification of repeated patterns in strings, trees and arrays. In *Fourth ACM Symposium on Theory of Computing*, pages 125–136, 1972.

[7] N. Pisanti, H. Soldano, and M. Carpentier. Incremental Inference of Relational Motifs with a Degenerate Alphabet. In *Combinatorial Pattern Matching (CPM)*, pages 229–240. Springer-Verlag, 2005. LNCS 3537.

[8] M.-F. Sagot. Spelling approximate repeated or common motifs using a suffix tree. In *Latin American Theoretical INformatics symposium (LATIN)*, pages 111–127. Springer-Verlag, 1998. LNCS 1380.

[9] H. Soldano, A. Viari, and M. Champesme. Searching for flexible repeated patterns using a non-transitive similarity relation. *Pattern Recognition Letters*, 16:243–246, 1995.

Nadia Pisanti
Laboratoire d'Informatique de Paris Nord
University of Paris 13, France
Email: pisanti@di.unipi.it

Henry Soldano
Laboratoire d'Informatique de Paris Nord
University of Paris 13, France
and
Atelier de Bioinformatique
University of Paris 6, France
Email: henry.soldano@lipn.univ-paris13.fr

Mathilde Capentier and Joel Pothier
Atelier de Bioinformatique
University of Paris 6, France

Site Consistency in Phylogenetic Networks with Recombination

TETSUO ASANO, PATRICIA EVANS, RYUHEI UEHARA, AND
GABRIEL VALIENTE

ABSTRACT. Perfect phylogeny is a fundamental model for the study of evolution. Given a set of binary sequences of the same length, perfect phylogeny is the problem of fitting the sequences as leaves of a rooted tree such that no site mutates more than once, and site consistency is the problem of finding a largest subset of sites that can fit a perfect phylogeny. The site consistency problem is known to be NP-hard, but polynomial-time solvable if the set of sequences can be derived on a particular form of phylogenetic network with recombination, called a galled-tree. In this paper, we introduce the problem dual to site consistency, called sequence consistency, provide a linear-time algorithm for site consistency when the set of sequences can be derived on a galled-tree, and establish fixed-parameter tractability of both site and sequence consistency.

1 Introduction

Perfect phylogeny is a fundamental model for the study of evolution. Consider a set of strings (sequences, in biological terms) of the same length, in matrix form. Rows stand for taxa (organisms, biological sequences, populations, languages) and columns for characters, whereas taxa are described by the states they exhibit on the set of characters. In the particular case of binary sequences, which is the focus of this paper, a character can be either absent or present on a given taxon.

Unlike the phylogenies studied in cladistics, though, the order of the characters on a sequence is assumed to be fixed. This is the case, for instance, of genomic sequences: rows stand for haplotypes, and sites in the haplotype represent single-nucleotide polymorphisms (single base substitutions of one nucleotide for another such that both extant and mutant alleles are observed in the population with enough frequency, above a certain threshold).

The evolutionary relationship between taxa is reflected in the absence or presence of particular characters in the binary sequences that represent them. For two taxa that differ on the state of a single character, the sequence in which that character is absent is an ancestor of the sequence in which the character is present, and their evolutionary relationship reflects in turn the emergence of that particular character over time. In such a case, there is a mutation of the sequence at the position or site corresponding to that character.

A phylogeny for a set M of n binary sequences of length m, is a rooted tree with exactly n leaves, each labeled by a distinct sequence of M, and with internal nodes labeled by ancestral sequences. All sequences are thus assumed to descend from a common ancestor, in which all characters are absent. Branches in the tree represent evolutionary relationships in the form of character state mutations, and it is assumed that no site mutates more than once in the whole tree.

The independent emergence of the same character along two different branches of a phylogeny, called homoplasy, is considered to be a poor indicator of evolutionary relationships, because similarity between the sequences does not reflect their shared ancestry [18], although there is recent evidence for extra-genomic inheritance of DNA sequence information [16]. A set of characters that admit a phylogeny without homoplasy, is said to be compatible and a phylogeny for a compatible set of characters is said to be a perfect phylogeny. In a perfect phylogeny, the set of nodes that exhibit a same character induce a subtree of the phylogeny [7].

The study of evolution in the presence of incompatible sets of characters can be made from two different perspectives: the model of evolution can be enriched, and the input set of binary sequences can be constrained. On the one hand, in an enriched model of evolution, mutation is complemented with (single-crossover) recombination, by which a prefix of one sequence is combined with a suffix of another sequence. The problem of finding a perfect phylogenetic network with the least possible number of recombination events is NP-hard [22], but it becomes polynomial-time solvable in a constrained form, called galled-tree [11], in which recombination cycles are edge-disjoint.

On the other hand, the input set of binary sequences can be constrained to a subset of the sequences or a subset of the sites. In systematic zoology, only the latter restriction has been studied so far, probably because cladistic characters are more often considered as hypotheses [14]. Site consistency is the problem of finding a largest subset of sites that can fit a perfect phylogeny, and it is known to be NP-hard [4], but polynomial-time solvable if the set of sequences can be derived on a galled-tree [10].

In this paper, we combine the two perspectives to the study of evolution in the presence of incompatible sets of characters. On the one hand, we introduce the problem dual to site consistency, called sequence consistency: the problem of finding a largest subset of sequences that can fit a perfect phylogeny. On the other hand, we provide a linear-time algorithm for a constrained form of both site and sequence consistency, and study the parameterized complexity of these problems.

The rest of the paper is organized as follows. Basic notions about sequence compatibility and perfect phylogenies are recalled in Section 1.1. Site consistency

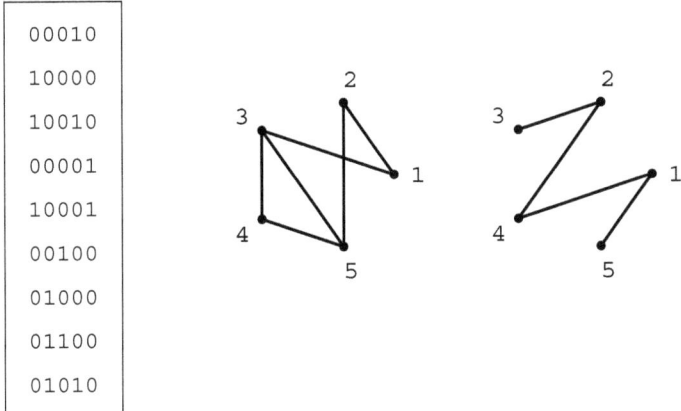

Figure 1. Compatibility graph (middle) and conflict graph (right) for an incompatible set of sequences (left).

is reviewed in Section 2 and sequence consistency is introduced in Section 3. The parameterized complexity of both site and sequence consistency is studied in Section 4. Finally, some conclusions are drawn in Section 5.

1.1 Preliminaries

Let us recall some basic notions about sequence compatibility and perfect phylogenies.

DEFINITION 2.1. Let M be a set of n binary sequences of length m. Two sites i and j, $1 \leq i, j \leq m$, are said to be *compatible* in M if M does not contain three sequences with the pairs $\{01, 10, 11\}$ in these two sites. Otherwise, the two sites are said to be *conflicting* in M. The *compatibility graph* for M is an undirected graph with a node for each site and an edge for each pair of compatible sites. The complement of the compatibility graph, with a node for each site and an edge for each pair of conflicting sites, is the *conflict graph* for M.

EXAMPLE 2.2. In the sequences shown in Figure 1, the first and fourth sites are in conflict: the first three sequences, 00010, 10000, 10010, become 01, 10, 11 when restricted to these sites. There are further incompatibilities among the sites in these sequences. Their whole compatibility graph and conflict graph are also shown in Figure 1.

In the absence of incompatibilities, the evolution of a set of binary sequences can be explained by means of a perfect phylogeny [2].

DEFINITION 2.3. Let M be a set of n binary sequences of length m. A *phy-

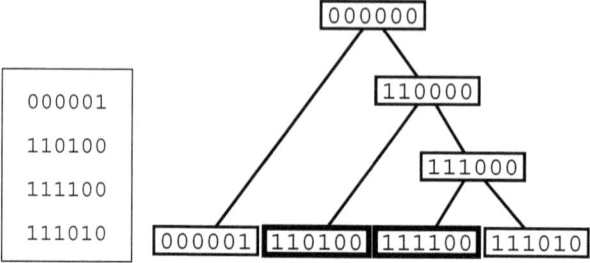

Figure 2. A set of sequences (left) that does not admit a perfect phylogeny. The phylogeny (right) has two nodes that exhibit the fourth site but are not connected by a path in the tree.

logeny for M is a rooted binary tree with exactly n leaves, each labeled by a distinct sequence of M, and with internal nodes labeled by hypothetical ancestors such that for each edge (X, Y) of T, the set of sites exhibited by X is contained in the set of sites exhibited by Y. A phylogeny T for M is called a *perfect phylogeny* if for each site i, $1 \leq i \leq n$, the set of nodes in T that exhibit the ith site induce a subtree of T.

EXAMPLE 2.4. The set of sequences shown in Figure 2 does not have a perfect phylogeny. In the phylogeny shown to the right, the set of nodes having a 1 in the fourth site, labeled by 110100 and 111100, are not connected by a path in the tree and thus, do not induce a subtree of the phylogeny.

2 Site Consistency

Simulations using stochastic models of population genetics, such as the Wright-Fischer neutral model [12], show that samples drawn from a population evolving according to an infinite-sites model of mutation (to be explained below) often have incompatible sites. In the presence of incompatibilities, the evolution of a set of binary sequences can no longer be explained by means of a perfect phylogeny.

The problem of inferring phylogenies on which the largest possible number of characters can be uniquely derived, has a long tradition in cladistics, dating back to [6, 14]. A cladistic character is said to be uniquely derived if it evolved only in one direction on a single occasion in its history. The uniquely-derived character concept clearly rules out homoplasy, the independent emergence of the same cladistic character along two different branches of a phylogeny.

In the context of genomic sequences, the uniquely-derived character concept is often referred to as the infinite-sites assumption, by which the m sites in a sequence are assumed to be so sparse relative to the mutation rate that in the time frame of interest, at most one mutation will have occurred at any site.

The site consistency problem was first formulated in [4].

DEFINITION 2.5. Let M be a set of n binary sequences of length m. The *site consistency problem* is to find the minimum number of sites to remove from M so that it can be derived on a perfect phylogeny.

The minimum number of sites to remove from a set M of n binary sequences of length m so that no conflict remains is given by a maximum clique of the compatibility graph for M or, alternatively, by a minimum node cover of the conflict graph for M. This is the essence of the NP-hardness proof of site consistency in [4].

THEOREM 2.6. *The site consistency problem is NP-hard.*

Given the NP-hardness of site consistency, it is natural to ask if the problem can be relaxed in some interesting and meaningful way. On the one hand, if there is a galled-tree explaining a given set of sequences, then the additional structure imposed by the galled-tree might be useful in solving the site consistency problem. On the other hand, if there is no galled-tree explaining a given set of sequences, then site consistency might be reformulated as the problem of finding a smallest number of sites that need to be removed in order for the set of sequences to have a galled-tree. Let us start with the latter reformulation of site consistency.

DEFINITION 2.7. Let M be a set of n binary sequences of length m. The *extended site consistency problem* is to find the minimum number of sites to remove from M so that it can be derived on a galled-tree.

Unfortunately, the problem of finding a smallest number of sites that need to be removed in order for the set of sequences to have a galled-tree remains NP-hard.

THEOREM 2.8. *The extended site consistency problem remains NP-hard.*

Proof. The solution is given by a maximum biconvex induced subgraph of the conflict graph. The class of convex graphs is closed under induced subgraphs, and node deletion problems are NP-complete for hereditary properties [15]. ■

Let us address now the former particular case of site consistency.

DEFINITION 2.9. Let M be a set of n binary sequences of length m that can be derived on a galled-tree. The *constrained site consistency problem* is to find the minimum number of sites to remove from M so that it can be derived on a perfect phylogeny.

The minimum number of sites to remove from a set M of binary sequences so that no conflict remains is given by a minimum node cover of the conflict graph for M and, if M can be derived on a galled-tree, then each non-trivial connected component of the conflict graph for M is biconvex and thus bipartite, and a minimum node cover can therefore be found in polynomial time [10]. By exploiting the additional structure of a biconvex conflict graph, we are able to find a minimum node cover in linear time, as sketched below. The proof will be published in full detail elsewhere.

THEOREM 2.10. *Given the conflict graph for a set of n binary sequences of length m, the constrained site consistency problem can be solved in $O(mn)$ time.*

Proof. Let M be a set of n binary sequences of length m that can be derived on a galled-tree, and let G be the conflict graph for M, which has m vertices and less than mn edges, because $m \leq 2n$ if M can be derived on a galled-tree [22]. Since G is biconvex, a maximum independent set of G, and thus a minimum node cover of G, can be obtained in $O(mn)$ time, as follows. If G is a bipartite permutation graph, with bipartition, say, (X, Y), then it can be partitioned into a sequence of complete bipartite graphs, say $K_1, K_2, \ldots, K_k$, and a maximum independent set of G can be obtained by maintaining for each $i = 1, 2, \ldots, k$ two candidates I_x^i and I_y^i for a maximum independent set of the subgraph of G induced by $K_1 \cup K_2 \cup \cdots \cup K_i$, where I_x^i is the candidate for maximum independent set that contains $X \cap K_i$ and I_y^i is the candidate for maximum independent set that contains $Y \cap K_i$. Then, the largest set of I_x^i and I_y^i for $i = k$ is a maximum independent set of G. Notice that it can be determined in $O(mn)$ time if G is a bipartite permutation graph [19].

Otherwise, if G is not a bipartite permutation graph, (X, Y) can be divided into the three bipartitions (X, Y_1), (X, Y_2), and (X, Y_3), where Y_1 and Y_3 contain all intervals that are linearly included, and Y_2 contains all proper (that is, not properly containing another one) intervals in the intersection model of G. Let G_j be the subgraph of G induced by (X, Y_j) for $j = 1, 2, 3$. Then, G_2 is a bipartite permutation graph, and a maximum independent set of G_2 can be obtained by the previous procedure. Moreover, G_1 and G_3 are chain graphs [13, 23], and they are also bipartite permutation graphs [20, Lemma 7]. Hence, maximum independent sets of G_1 and G_3 can also be obtained by the previous procedure. By merging these three independent sets, a maximum independent set of G can be obtained in $O(mn)$ time. ∎

The conflict graph itself can be computed in $O(m^2 n)$ time [1, 11]. However, if there is a galled-tree for M, then it was established in [22] that m can be at most $2n$ and thus, $O(m^2 + mn) = O(mn)$. Therefore, an improved algorithm for computing the conflict graph in $O(m^2 + mn)$ time would yield a solution to the constrained site consistency problem in $O(mn)$ time. By using a compact representation of the intervals of zero and one sequences for each site, we are able to compute the conflict graph in $O(m^2 + mn)$ time, as sketched below. The proof will be published in full detail elsewhere.

LEMMA 2.11. *Let M be a set of n binary sequences of length m. The conflict graph for M can be computed in $O(m^2 + mn)$ time.*

Proof. Sort M in row order in $O(mn)$ time, by radix sorting techniques [21, App. A.5]. Define for each site $1 \leq j \leq m$, the ordered list 0_j of maximal intervals $[0_\ell, 0_r]$ such that $M_{i,j} = 0$ for all $\ell \leq i \leq r$, and the ordered list 1_j of maximal intervals $[1_\ell, 1_r]$ such that $M_{i,j} = 1$ for all $\ell \leq i \leq r$.

It is easy to see that the lists of zero and one intervals defined above can be constructed in $O(mn)$ time, by traversing M in column order.

Once the lists of zero and one intervals are available, we can build the conflict graph in $O(m^2 + mn)$ time as follows. The idea is to enumerate the set P_{ab} of pairs (p_i, p_j) for which there exists a sequence s_k such that $M(s_k, p_i) = a$ and $M(s_k, p_j) = b$. Then, the set of all pairs of conflicting sites is given by the intersection of the three sets $P_{01} \cap P_{10} \cap P_{11}$.

The set P_{01} can be computed by performing, for each pair i, j of sites with $1 \leq i, j \leq m$ and $i \neq j$, a simultaneous traversal of 0_i and 1_j during which, upon intervals $[0_\ell, 0_r]$ of 0_i and $[1_\ell, 1_r]$ of 1_j, we advance along 0_i if $0_r < 1_\ell$, advance along 1_j if $1_r < 0_\ell$, and otherwise, if $0_\ell \leq 1_\ell \leq 0_r$ or $1_\ell \leq 0_\ell \leq 1_r$, there exists a sequence s_k such that $M(s_k, p_i) = 0$ and $M(s_k, p_j) = 1$. The same procedure allows us to compute the sets P_{10} and P_{11}.

The total number of interval comparisons performed for computing the sets P_{01}, P_{10}, and P_{11} is bounded by $O(mn)$. This guarantees that the algorithm runs in $O(m^2 + mn)$ time. ∎

Now, if the given set of sequences can be derived on a galled-tree, Theorem 2.10 and Lemma 2.11 entail that the constrained site consistency problem can be solved in time linear in the size of the set of sequences.

COROLLARY 2.12. *Given a set of n binary sequences of length m that can be derived on a galled-tree, the constrained site consistency problem can be solved in $O(mn)$ time.*

While it is NP-hard, the site consistency problem is fixed-parameter tractable; this is discussed in Section 4.

3 Sequence Consistency

In the presence of incompatibilities, evolution can be studied by constraining the input set of binary sequences to either a subset of the sites or a subset of the sequences. The former was discussed in the previous section. Let us focus now on the latter.

DEFINITION 2.13. Let M be a set of n binary sequences of length m. Three sequences i, j and k, $1 \leq i, j, k \leq n$, are said to be *compatible* in M if M does not contain two sites with the triplets $\{011, 101\}$ in these two sites. Otherwise, the three sequences are said to be *conflicting* in M. The *compatibility hypergraph* for M is a hypergraph with a node for each sequence and a hyperedge for each triplet

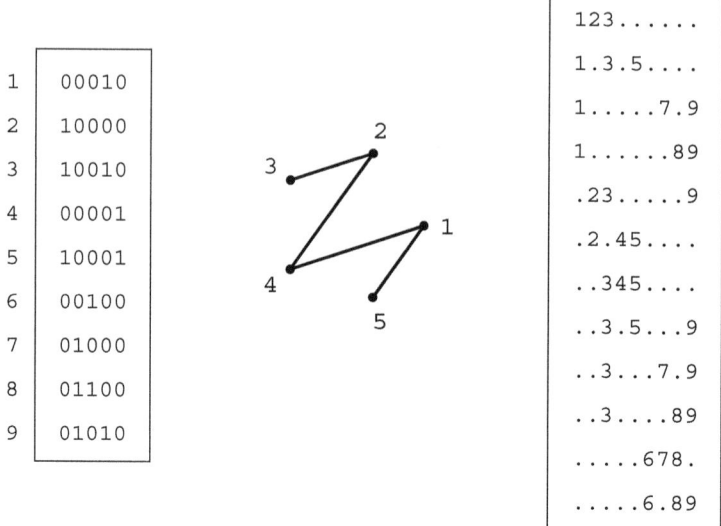

Figure 3. Conflict graph (middle) and conflict hypergraph (right) for the incompatible set of sequences (left) of Figure 1.

of compatible sequences. The complement of the compatibility hypergraph, with a node for each sequence and a hyperedge for each triplet of conflicting sequences, is the *conflict hypergraph* for M.

EXAMPLE 2.14. In the sequences shown in Figure 3, the first three sequences are in conflict: the first and fourth sites, 011010000 and 101000001, become 011 and 101 when restricted to these sequences. There are further incompatibilities among these data. Their whole conflict hypergraph is also shown in Figure 3.

Parallel to site consistency is the sequence consistency problem.

DEFINITION 2.15. Let M be a set of n binary sequences of length m. The *sequence consistency problem* is to find the minimum number of sequences to remove from M so that it can be derived on a perfect phylogeny.

The minimum number of sequences to remove from a set M of n binary sequences of length m so that no conflict remains, is given by a minimum hitting set of the conflict hypergraph for M. This is the essence of the following proof, which extends the NP-hardness proof of site consistency in [4] to sequence consistency.

THEOREM 2.16. *The sequence consistency problem is NP-hard.*

Proof. Recall that the NP-complete HITTING SET problem is to determine, given a collection C of subsets of a finite set S and a positive integer $K \leq |S|$, if there is a subset $S' \subseteq S$ with $|S'| \leq K$ such that S' contains at least one element from each subset $c \in C$, and that HITTING SET remains NP-complete even if $|c| \leq 2$ for all $c \in C$ [9].

Let (M, m, n) be an instance of the sequence consistency problem. A corresponding instance (S, C) of HITTING SET with $|c| = 3$ for all $c \in C$ can be obtained as follows. Let $S = \{1, \ldots, n\}$, and let $(i, j, k) \in C$ if and only if $M(i, x) = 0$, $M(i, y) = 1$, $M(j, x) = 1$, $M(j, y) = 0$, $M(k, x) = 1$, and $M(k, y) = 1$ for some $1 \leq x \leq y \leq m$.

Conversely, let (S, C) be an instance of HITTING SET with $|c| = 3$ for all $c \in C$. A corresponding instance (M, m, n) of the sequence consistency problem can be obtained as follows. Let M be a binary matrix with $n = |S|$ rows and $m = 2|C|$ columns. Set all entries of M to 0. For each subset $(i, j, k) \in C$, let x, y be the corresponding columns in M, and set entries $M(i, y)$, $M(j, x)$, $M(k, x)$, and $M(k, y)$ to 1.

Then, $(i, j, k) \in C$ if and only if rows i, j, k are incompatible and therefore, there is a subset $S' \subseteq S$ with $|S'| \leq K$ such that S' contains at least one element from each subset $c \in C$ if and only if M has a compatible subset of rows of size at least $n - K$. ■

Since 3-hitting set is fixed-parameter tractable [8, 17] this reduction does not apply to the parameterization of sequence consistency; in Section 4 we show that sequence consistency is fixed-parameter tractable.

4 Parameterized Complexity

The theory of parameterized complexity [5] enables us to ask questions about the difficulty of solving problems when certain problem parameters are kept small, and the solution is allowed to run in time exponential in those parameters while still polynomial in the overall input size. For the problems of site consistency and sequence consistency, we investigate the complexity for instances that are close to being a perfect phylogeny, that is, if the number of sites or sequences that must be deleted is restricted by a given parameter.

Since the site consistency problem for k site deletions can be solved by finding a node cover of size k in the conflict graph, it is fixed-parameter tractable [5] and can be solved in $O(m^2 + mn + 1.2738^k)$ time by constructing the conflict graph in $O(m^2 + mn)$ time as in Lemma 2.11 and then finding a node cover in $O(1.2738^k + km)$ time [3]. The km term disappears since $k \leq m \leq m^2$.

Sequence consistency for k sequence deletions can also be examined from the parameterized perspective, and is also fixed-parameter tractable despite being NP-hard.

THEOREM 2.17. *Sequence consistency is fixed-parameter tractable, and can be solved in $O(3^k m^2 n)$ time where k is the number of sequences to be removed.*

Proof. The remaining rows will have a perfect phylogeny if and only if there are no pairs of columns that contain all of $\{10, 01, 11\}$. So, for each pair of columns x, y we define

$$\begin{aligned} L_{x,y} &= \{i \mid \text{row } i \text{ has 1 in column } x \text{ and 0 in column } y\} \\ R_{x,y} &= \{i \mid \text{row } i \text{ has 0 in column } x \text{ and 1 in column } y\} \\ I_{x,y} &= \{i \mid \text{row } i \text{ has 1 in column } x \text{ and 1 in column } y\} \end{aligned}$$

If $L_{x,y}$, $R_{x,y}$, and $I_{x,y}$ are all nonempty, then we have a conflict and thus cannot have a perfect phylogeny. So for any such pair of columns x, y, all rows of at least one of the three sets must be deleted from M.

Since we do not know which of the sets it is best to delete, we will try all combinations; each such deletion will decrease the value of k by at least one, so we have a search tree of branch factor 3 whose depth is bounded by k.

SeqCons(S, k)
if $k > 0$ **then**
    pick any x, y where $L_{x,y}$, $R_{x,y}$, and $I_{x,y}$ all $\neq \emptyset$
    return SeqCons($S \setminus L_{x,y}, k - |L_{x,y}|$) $\vee$ SeqCons($S \setminus R_{x,y}, k - |R_{x,y}|$) $\vee$ SeqCons($S \setminus I_{x,y}, k - |I_{x,y}|$)
else
    if there are still x, y where $L_{x,y}$, $R_{x,y}$, and $I_{x,y}$ all $\neq \emptyset$ **then**
        return false
    else
        return true

Each set removal can be used to update the contents of the other sets in $O(m^2 n)$ time, so this algorithm runs in $O(3^k m^2 n)$ time and shows the fixed-parameter tractability of sequence consistency. ∎

5 Conclusion

The site consistency problem is known to be NP-hard, but polynomial-time solvable if the given set of sequences can be derived on a particular form of phylogenetic network with recombination, called a galled-tree. For a set of n binary sequences of length m that can be derived on a galled-tree, we build the conflict graph in $O(m^2 + mn)$ time and find a minimum node cover of the conflict graph in $O(mn)$ time, thus solving the site consistency problem in $O(mn)$ time, because

$m \leq 2n$ if there is a galled-tree for the given set of sequences [22]. We thus answer in the affirmative a conjecture of D. Gusfield [10, Section 4.1]. We also introduce the problem dual to site consistency, called sequence consistency, show that site and sequence consistency are both fixed-parameter tractable, solve the site consistency problem in $O(m^2 + mn + 1.2738^k)$ time, where k is the number of site deletions, and solve the sequence consistency problem in $O(3^k m^2 n)$ time, where k is the number of sequence deletions.

Acknowledgments

The last author would like to thank Dan Gusfield for introducing him to the problem of site consistency in galled-trees, and Francesc Rosselló for useful discussions about the topic of this paper. G. Valiente was partially supported by the Japan Society for the Promotion of Science through Long-term Invitation Fellowship L-05511 for visiting JAIST (Japan Advanced Institute of Science and Technology).

BIBLIOGRAPHY

[1] V. Bafna and V. Bansal. The number of recombination events in a sample history: Conflict graph and lower bounds. *IEEE Transactions on Computational Biology and Bioinformatics*, 1(2):78–90, 2004.

[2] H. L. Bodlaender, M. R. Fellows, and T. J. Warnow. Two strikes against perfect phylogeny. In *Proc. 19th Int. Coll. Automata, Languages and Programming*, volume 623 of *Lecture Notes in Computer Science*, pages 273–283. Springer-Verlag, 1992.

[3] J. Chen, I. A. Kanj, and G. Xia. Simplicity is beauty: Improved upper bounds for vertex cover. Technical Report TR05-008, DePaul University School of Computer Science, Telecommunications and Information Systems, April 2005.

[4] W. H. E. Day and D. Sankoff. Computational complexity of inferring phylogenies by compatibility. *Systematic Zoology*, 35(2):224–229, 1986.

[5] R. G. Downey and M. R. Fellows. *Parameterized Complexity*. Springer-Verlag, New York, 1999.

[6] G. F. Estabrook, J. C. S. Johnson, and F. R. McMorris. An algebraic analysis of cladistic characters. *Discrete Mathematics*, 16(2):141–147, 1976.

[7] D. Fernández-Baca. The perfect phylogeny problem. In X. Cheng and D.-Z. Du, editors, *Steiner Trees in Industry*, volume 11 of *Combinatorial Optimization*, pages 203–234. Springer-Verlag, 2002.

[8] H. Fernau. A top-down approach to search-trees: Improved algorithms for 3-hitting set. *Electronic Colloquium on Computational Complexity*, 11(73):1–31, 2004.

[9] M. R. Garey and D. S. Johnson. *Computers and intractability: A guide to NP-completeness*. Freeman, 1979.

[10] D. Gusfield, S. Eddhu, and C. Langley. The fine structure of galls in phylogenetic networks. *INFORMS Journal on Computing*, 16(4):459–469, 2004.

[11] D. Gusfield, S. Eddhu, and C. Langley. Optimal, efficient reconstruction of phylogenetic networks with constrained recombination. *Journal of Bioinformatics and Computational Biology*, 2(1):173–213, 2004.

[12] R. R. Hudson. Generating samples under a Wright-Fisher neutral model of genetic variation. *Bioinformatics*, 18(2):337–338, 2002.

[13] T. Kloks, D. Kratsch, and H. Müller. Bandwidth of Chain Graphs. *Information Processing Letters*, 68(6):313–315, 1998.

[14] W. J. Le Quesne. A method of selection of characters in numerical taxonomy. *Systematic Zoology*, 18(1):201–205, 1969.

[15] J. M. Lewis and M. Yannakakis. The node-deletion problem for hereditary properties is NP-complete. *Journal of Computer and System Sciences*, 20(2):219–230, 1980.
[16] S. J. Lolle, J. L. Victor, J. M. Young, and R. E. Pruitt. Genome-wide non-mendelian inheritance of extra-genomic information in Arabidopsis. *Nature*, 434(7032):505–509, 2005.
[17] R. Niedermeier and P. Rossmanith. An efficient fixed-parameter algorithm for 3-hitting set. *Journal of Discrete Algorithms*, 1(1):89–102, 2003.
[18] R. D. M. Page and E. C. Holmes. *Molecular Evolution: A Phylogenetic Approach*. Blackwell Science, 1998.
[19] J. Spinrad, A. Brandstädt, and L. K. Stewart. Bipartite permutation graphs. *Discrete Applied Mathematics*, 18(1):279–292, 1987.
[20] R. Uehara and Y. Uno. Efficient Algorithms for the Longest Path Problem. In *Proc. 15th Ann. Int. Symp. Algorithms and Computation*, volume 3341 of *Lecture Notes in Computer Science*, pages 871–883. Springer-Verlag, 2004.
[21] G. Valiente. *Algorithms on Trees and Graphs*. Springer-Verlag, Berlin, 2002.
[22] L. Wang, K. Zhang, and L. Zhang. Perfect phylogenetic networks with recombination. *Journal of Computational Biology*, 8(1):69–78, 2001.
[23] M. Yannakakis. Node-Deletion Problems on Bipartite Graphs. *SIAM Journal on Computing*, 10(2):310–327, 1981.

Tetsuo Asano and Ryuhei Uehara
Japan Advanced Institute of Science and Technology
School of Information Science
Asahidai 1-1, Nomi, Ishikawa 923-1292, Japan
Email: t-asano@jaist.ac.jp and uehara@jaist.ac.jp

Patricia A. Evans
Faculty of Computer Science
University of New Brunswick
P. O. Box 4400, Fredericton, NB, Canada E3B 5A3
Email: pevans@unb.ca

Gabriel Valiente
Department of Software
Technical University of Catalonia
E-08034 Barcelona, Spain
Email: valiente@lsi.upc.edu

Classification of Splice-junction Gene Sequences by a Special Type of Threshold Circuits

GEORGIOS LAPPAS, KATHLEEN STEINHÖFEL, AND
ANDREAS A. ALBRECHT

ABSTRACT. We investigate the splice-junction gene sequences database (SJGSD) from the UCI Machine Learning Repository in the context of threshold circuit complexity, i.e. we attempt to find some rule that allows us to calculate *a priori* the number of threshold gates that is sufficient to achieve a small error rate after training a circuit on sample data. The particular threshold gates are computed by a combination of the classical perceptron algorithm with a specific type of stochastic local search. The circuit complexity is analysed for depth-two and depth-four threshold circuits, where we introduce a novel approach to compute depth-four circuits. For the SJGSD we obtain approximately the same size of depth-two and depth-four circuits for the best classification rates on test samples. Based on classical results from threshold circuit theory and our experimental observations we suggest a simple formula to calculate an upper bound for the number of threshold gates that provides a high degree of generalisation of sample data.

Keywords: Machine learning; circuit complexity; simulated annealing.

1 Introduction

The present paper continues the research outlined in [4] on the threshold circuit complexity of classification problems. Usually, the application of a particular learning-based classification tool to a specific problem requires a certain amount of adjustment, which is apparently unavoidable in the light of the *No Free Lunch Theorems* [33]. In our approach, we try to establish a set of simple rules that allows us to calculate estimations of problem-dependent parameters. Our learning heuristic basically combines the classical perceptron algorithm [25] with a specific type of stochastic local search [14] for finding threshold gates of a classification circuit, where only a few parameters have to be adjusted to a given problem [3, 4].

We specifically investigate the circuit complexity inherent to a given problem, i.e. we attempt to find *a priori* estimations of the number of threshold gates that ensure a high degree of generalisation of sample data. The number of threshold gates is analysed for two types of threshold circuits, namely depth-two and depth-four circuits. Both types of circuits are calculated by the above-mentioned learning procedure with respect to single threshold gates, where we introduce a novel approach to compute depth-four circuits. The approach is applicable to circuits of any even depth $d = 2 \cdot j$; however, the circuit size increases exponentially in d.

For circuits of depth $d \geq 2$ we proceed as follows: A given sample set $\mathcal{S}$ is split into two disjoint subsets $\mathcal{S}_\mathrm{L}$ and $\mathcal{S}_\mathrm{T}$, where $\mathcal{S}_\mathrm{L}$ with $\mid \mathcal{S}_\mathrm{L} \mid \approx \frac{2}{3} \cdot \mid \mathcal{S} \mid$ samples is used to calculate the approximating threshold circuit, and the rest of the samples $\mathcal{S}_\mathrm{T}$ is left for testing purposes. The set $\mathcal{S}_\mathrm{L}$ is then further partitioned into $d/2$ subsets $\mathcal{S}_\mathrm{L}^j$ of approximately the same size, $1 \leq j \leq d/2$, and we apply the following recursive procedure: $\mathcal{S}_\mathrm{L}^1$ is used to calculate N_1 threshold circuits $\mathcal{C}_1$ of depth 2, each consisting of t perceptrons directly connected to the input data. The t gates of a single subcircuit $\mathcal{C}_1$ are combined by a voting function with a pre-determined threshold. Then, if the N_{j-1} circuits $\mathcal{C}_{j-1}$ of depth $2 \cdot (j-1)$ have been calculated already, we generate from $\mathcal{S}_\mathrm{L}^j$ randomly N_j primary training sets $T^{\mathrm{pr}}_{[j,i]}$, $1 \leq i \leq N_j$, and each $T^{\mathrm{pr}}_{[j,i]}$ applied to circuits of type $\mathcal{C}_{j-1}$ generates a secondary training set $T^{\mathrm{snd}}_{[j,i]}$. The sets $T^{\mathrm{snd}}_{[j,i]}$ are then used together with the classification information from $\mathcal{S}_\mathrm{L}^j$ to calculate the perceptron gates at depth $2 \cdot j - 1$, which are then again combined by a voting function at depth $2 \cdot j$.

We tested the approach on the splice-junction gene sequences database from the UCI Machine Learning Repository [30] that, actually, defines three classification problems. By $\mathcal{C}_d^P[e = R]$ we denote circuits that achieve an error rate of R on test samples of problem P. We compare the "typical" gate complexity in circuits $\mathcal{C}_2^P[e = R]$ and $\mathcal{C}_4^P[e = R]$ for R close to the lowest error rates we obtain on problems P. The effect of circuit depth on circuit size is one of the hardest problems in theoretical computer science. The problem emerged already in the discussion about perceptrons [23] in the context of the circuit complexity of $XOR(x_1, ..., x_n)$. To identify sequences of Boolean functions $\{f(x_1, ..., x_n)\}_{n=n_0}^{\infty}$ with "superpolynomial" (or exponential) gate number in constant depth circuits of unbounded fan-in gates is a very difficult problem and only slow progress has been made over the past decades. For example, in [24] a sequence of functions was designed that requires at least the superpolynomial number $n^{\Omega(\log n)}$ of threshold gates in any circuit of depth 3. Furthermore, it has been shown [5, 10] that computing the permanent of an $n \times n$ matrix requires a superpolynomial number of gates in any threshold circuit of constant depth; for more information about the computational power of small depth circuits we refer the reader to [32].

Since it seems to be very difficult to find sequences of complex, explicitly defined functions even for small values of d, one can ask whether the inherent complexity of datasets from real-world problems is actually relatively low, i.e. if the datasets admit circuit representations of low complexity for a sufficiently large circuit depth $d > 2$ and unbounded fan-in gates. We note that here we are dealing with particular functions of fixed variable numbers, not with sequences of functions, and our $\mathcal{C}_d$ circuits cover only a small range of all possible depth d circuits of unbounded fan-in linear threshold gates. Nevertheless, the comparison of $\mathcal{C}_2^P[e = R]$ and $\mathcal{C}_4^P[e = R]$ provides some empirical evidence that the best classification rate is obtained on approximately the same circuit size, i.e. the same number of linear threshold functions. Moreover, counting the gates for the asymptotically optimal design of linear threshold circuits for the most complex n-ary Boolean functions [19, 21] results in the same range of circuit size as in $\mathcal{C}_2^P[e \approx R_{\min}]$ and $\mathcal{C}_4^P[e \approx R_{\min}]$ for all three classification problems derived from SJGSD (the actual input size n_L^P is taken from the number of bits necessary to enumerate all training samples of P). Based on this observation, we propose a formula to estimate the number of perceptrons that have to be trained in order to achieve a high classification accuracy. Of course, for problems of low complexity or problems that are even linearly separable, like the Mushroom Dataset from the UCI Machine Learning Repository [30], the proposed formula overestimates the number of threshold gates significantly.

2 The Method

In our approach, we combine the perceptron algorithm with stochastic local search. The research on perceptron algorithms has a long history and goes along with the efforts to find solutions for systems of linear inequalities $l^j(\vec{z}) = \vec{a}^j \cdot \vec{z} + b^j \geq 0$, $j = 1, \ldots, m$. S. AGMON [2] proposed a simple iteration procedure that starts with an arbitrary initial vector $\vec{z}_0$. When $\vec{z}_i$ does not represent a solution of the system, then $\vec{z}_{i+1}$ is taken as the orthogonal projection of the farthest hyperplane which corresponds to a violated linear inequality: $\vec{z}_{i+1} := \vec{z}_i + t \cdot \vec{a}^{j_0}$, where $t = -l^{j_0}(\vec{z}_i)/ \mid \vec{a}^{j_0} \mid^2$ and $\vec{a}^{j_0}$ maximizes $-l^j(\vec{z}_i)/ \mid \vec{a}^j \mid^2$ among the violated $l^j(\vec{z}_i)$. The procedure is similar to the so-called relaxation method which was developed by G. TEMPLE [28] for finding solutions of linear equations.

2.1 The Perceptron Algorithm

S. AGMON's method later became known as the classical perceptron algorithm [25]. If the set of points $\vec{a}^j$, $j = 1, \ldots, m$ and $\vec{a}^j = (a_1^j, \ldots, a_n^j)$, can be separated by a linear function, the following convergence property can be proved for the perceptron algorithm [23]: Let $S = \{\vec{a}^j\}$ denote the "sample set" of input vectors classified as positive and negative samples. If $\vec{w}^*$ is a unit vector solution to the separation problem, i.e., $\vec{w}^*\vec{a} > 0$ for all $[\vec{a}, +] \in S$ and $\vec{w}^*\vec{a} < 0$ for all $[\vec{x}, -] \in S$, then the perceptron algorithm converges in at most $1/\sigma^2$ iterations, where $\sigma := \min_{[\vec{a}, \eta] \in S} \mid \vec{w}^* \cdot \vec{a} \mid$, $\eta \in \{+, -\}$. The parameter σ has the interpretation of $\cos(\vec{w}^*, \vec{a})$ for the angle between $\vec{w}^*$ and $\vec{a}$. The value of σ can be exponentially small in terms of the dimension n.

In general, the simple perceptron algorithm performs well even if the sample set is not consistent with any weight vector $\vec{w}$ of linear threshold functions [13, 27]. If the sample set is linearly separable, BAUM [6] has shown that under modest assumptions it is likely that the perceptron algorithm will find a highly accurate approximation of a solution vector $\vec{w}^*$ in polynomial time.

On the other hand, HÖFFGEN and SIMON [15] proved that finding a linear threshold function that minimises the number of misclassified vectors $\vec{a}^j$ is NP-hard for arbitrary sample sets (see also [16] for sigmoid functions). Variants of the perceptron algorithm on sample sets that are inconsistent with linear separation are presented in [7, 9]. For example, if the (average) inconsistency with linear separation is small relative to σ, then with high probability the perceptron algorithm will achieve a good classification of samples in polynomial time [9].

In our approach, we consider an extension of the perceptron algorithm by a simulated annealing-based search strategy [1, 11, 14, 17]. The simulated annealing procedure employs a logarithmic cooling schedule $c(k) = \Gamma/\ln(k+2)$ where the "temperature" decreases at each step.

2.2 Logarithmic Simulated Annealing

The simulated annealing-based extension of the perceptron algorithm is activated when the number of misclassified examples increases for the new hypothesis compared to the previous one. In this case, a random decision is made according to the rules of simulated annealing. If the new hypothesis is rejected, a random choice is made among the misclassified examples for the calculation of the next hypothesis. To describe our extension of the perceptron algorithm in more details, we have to define the configuration space together with a neighborhood relation.

Configuration Space and Neighbourhood Relation

The configuration space consists of all linear threshold functions with rational weights w_i represented by pairs of binary tuples each of length r: $w_i \in (\pm 1) \cdot \{0,1\}^r \times \{0,1\}^r$, $i = 1, ..., n$. We assume that the elements of the configuration space do have a fixed n^{th} coordinate, i.e. w_n represents the threshold. We denote the configuration space by

$$\mathcal{F} = \{ f(\vec{x}) : f(\vec{x}) = \sum_{i=1}^{n} w_i \cdot x_i, \ w_i \in (\pm 1) \cdot \{0,1\}^r \times \{0,1\}^r \}. \quad (1)$$

The neighborhood relation depends on the given sample set S, where

$$S = \{ [\vec{a}, \eta] : \vec{a} = (a_1, ..., a_n), \ a_n = 1, \ a_i = (p_i, q_i), \ p_i, q_i \in \{0,1\}^r, \ \eta \in \{+,-\} \}. \quad (2)$$

We define the objective function through the set of examples misclassified by the current configuration $f(\vec{x})$:

$$S\Delta f(\vec{a}) := \{ [\vec{a}, \eta] : \vec{a} \in \mathcal{S} \text{ and } f(\vec{a}) < 0 \,\&\, \eta = + \text{ or } f(\vec{a}) \geq 0 \,\&\, \eta = - \}. \tag{3}$$

The objective function is given by

$$\mathcal{Z}(f(\vec{x})) := |S\Delta f(\vec{x})|. \tag{4}$$

The set $\mathcal{N}_f$ of potential neighbours of $f(\vec{x})$ is derived from $S\Delta f(\vec{x})$ in accordance with the perceptron algorithm:

$$\mathcal{N}_f := \left\{ f' \;\middle|\; w'_i := w_i - \frac{\sum_{i=1}^n w_i \cdot a_i}{\sqrt{\sum_{i=1}^n a_i^2}} \cdot a_i, \; \vec{a} \in S\Delta f \right\} \cup \{f\}. \tag{5}$$

The probability of performing the transition between f and f' is defined by

$$\mathbf{Pr}\{f \to f'\} = \begin{cases} G[f, f'] \cdot A[f, f'], & \text{if } f' \neq f, \\ 1 - \sum_{g \neq f} G[f, g] \cdot A[f, g], & \text{otherwise}, \end{cases} \tag{6}$$

where $G[f, f']$ denotes the generation probability and $A[f, f']$ is the probability of accepting f' once it has been generated by f.

Generation and Acceptance Probabilities

We use a non-uniform generation probability which is based upon $S\Delta f$ from (3): For $f' \in \mathcal{N}_f$ associated with $\vec{a} \in S\Delta f$ in (5) we set

$$U(\vec{a}) := \begin{cases} |f(\vec{a})|, & \text{if } f(\vec{a}) < 0 \text{ and } \eta(\vec{a}) = +, \\ f(\vec{a}), & \text{if } f(\vec{a}) \geq 0 \text{ and } \eta(\vec{a}) = -. \end{cases} \tag{7}$$

The generation probability is then defined by

$$G[f, f'] := \frac{U(\vec{a})}{\sum_{\vec{a} \in S\Delta f} U(\vec{a})}. \tag{8}$$

Thus, preference is given to the neighbours that maximise the deviation. The acceptance probabilities $A[f, f']$, $f' \in \mathcal{N}_f$, are derived from the underlying analogy to thermodynamic systems:

$$A[f, f'] := \begin{cases} 1, & \text{if } \mathcal{Z}(f') - \mathcal{Z}(f) \leq 0, \\ e^{-(\mathcal{Z}(f') - \mathcal{Z}(f))/c}, & \text{otherwise,} \end{cases} \quad (9)$$

where c is a control parameter having the interpretation of a *temperature* in annealing procedures. The actual decision, whether or not f' should be accepted for $\mathcal{Z}(f') > \mathcal{Z}(f)$, is performed in the following way: f' is accepted, if

$$e^{-(\mathcal{Z}(f') - \mathcal{Z}(f))/c} \geq \rho, \quad (10)$$

where $\rho \in [0, 1]$ is a uniformly distributed random number. The value ρ is generated in each trial in case of $\mathcal{Z}(f') > \mathcal{Z}(f)$.

Inhomogeneous Markov Chains

Let $\mathbf{a}_f(k)$ denote the probability of being in configuration $f \in \mathcal{F}$ after k steps have been performed according to (6),...,(10). The probability $\mathbf{a}_f(k)$ is given by

$$\mathbf{a}_f(k) := \sum_h \mathbf{a}_h(k-1) \cdot \mathbf{Pr}\{h \to f\}, \quad (11)$$

where $\mathbf{Pr}\{h \to f\}$ is from (6). The recursive application of (11) defines a Markov chain of probabilities $\mathbf{a}_f(k)$, where $f \in \mathcal{F}$ and $k = 1, 2, \ldots$. If the parameter $c = c(k)$ in (9) is a constant c, the chain is said to be a *homogeneous* Markov chain; otherwise, if $c(k)$ is lowered at any step, the sequence of probability vectors $\vec{\mathbf{a}}(k)$ is an *inhomogeneous* Markov chain.

We consider a special type of inhomogeneous Markov chains only. The motivation for this choice is based upon the convergence properties of the two types of Markov chains: Convergence propositions about homogeneous Markov chains rely on an infinite number of transitions at fixed "temperatures" c. The probability distribution approached in the limit is the Boltzmann distribution $e^{-\mathcal{Z}(f)/c}/F$, where F is a normalisation value. If $c \to 0$, the Boltzmann distribution tends to the distribution over optimum configurations. In practice, however, it is infeasible to perform an infinite number of transitions at fixed temperatures. The convergence analysis of inhomogeneous Markov chains avoids the intermediate step, and in our approach the "temperature" $c(k)$ changes in accordance with

$$c(k) = \frac{\Gamma}{\ln(k+2)}, \quad k = 0, 1, \ldots. \quad (12)$$

The choice of $c(k)$ is motivated by HAJEK's theorem [14] on logarithmic cooling schedules. We denote by $\mathcal{F}_{\min}$ the set of optimum solutions (minimising the classification error). Basically, HAJEK's theorem states

THEOREM 3.1. *Under some natural assumptions about $\mathcal{F}$ and the neighbourhood relation $\mathcal{N}_f$, the asymptotic convergence $\sum_{f \in \mathcal{F}_{\min}} \mathbf{a}_f(k) \xrightarrow[k \to \infty]{} 1$ of the stochastic algorithm defined by (6),...,(10) is guaranteed if and only if Γ from (12) is lower bounded by the maximum value of the minimum escape height from local minima.*

Thus, the speed of convergence associated with the logarithmic cooling schedule (12) is mainly defined by the value of Γ. Since we deal with sample data and a relatively complicated neighbourhood relation (5), we cannot verify the applicability of Theorem 3.1. Nevertheless, previous research [3, 4] encouraged us to employ (12) in our stochastic local search procedure.

2.3 The Generation of Classification Circuits

The stochastic procedure described in Section 2.2 computes a single threshold function (gate). The number of transitions $k \leq K$ actually performed is a parameter of our approach (if $S\Delta f = \emptyset$ for $k < K$, the procedure is terminated, of course).

It is important to note that we aim at problems that are not linearly separable, even for relatively small subsets of the sample set, which are used to train single threshold gates of the circuit (otherwise, the "pure" perceptron algorithm would be more appropriate). Therefore, we don't try, in general, to achieve zero classification error on training data.

The classification circuit itself can be described by an inductive procedure as already outlined in Section 1. Apart from Γ and K, we choose some more parameters: $p, q, t, \alpha \geq 1$, $t/2 \leq \vartheta \leq t$, and $d \geq 2$, where d is an even number. Furthermore, we assume that a basic sample set $\mathcal{S} = \{[\vec{a}, \eta], \eta \in \{+, -\}\}$ of positive and negative examples is given, where the number of positive and negative samples is "balanced".

The Structure of Classification Circuits

We use a homogeneous structure of circuits consisting of at most three types of gates: perceptrons calculated from $\mathcal{S}$ (see Section 2.2 and Section 2.3), fixed type voting functions, and fixed type counting functions. The structure of classification circuits $\mathcal{C}_d$ is described by induction:

1. $d = 2$: $\mathcal{C}_2$ consists of t perceptrons, each of them connected to the $(n-1)$ input variables. The binary outputs of the t threshold gates (perceptrons) are the inputs to a simple voting function $\sum_{u=1}^{t} y_u \geq \vartheta$; see Figure 1. The threshold $\vartheta \geq t/2$ is a parameter of our approach.

2. $d = 2 \cdot j > 2$: $\mathcal{C}_d$ is built from $q \cdot t$ modified copies of circuits of type $\mathcal{C}_{d-2}$. The modification relates to the output gate of circuits of type $\mathcal{C}_{d-2}$: The voting function is substituted by a counting function $\sum_{u=1}^{t} y_u$, i.e. the

outputs are now integers (rather than binary values only), and the circuits are denoted by $\mathcal{C}'_{d-2}$. We take q copies of type $\mathcal{C}'_{d-2}$ to constitute the inputs to a single threshold function (perceptron). The t binary outputs of the threshold gates (each depending on q variable inputs) are again the inputs to the voting function $\sum_{u=1}^{t} y_u \geq \vartheta$; see Figure 2.

We note that perceptrons are "located" at depths $d' = 2 \cdot j - 1$ only, $j \geq 1$. By simple calculations we obtain the following formula for the number $N(d, d')$, $d \geq d' + 1$, of perceptrons at depth $d' = 2 \cdot j - 1$ in a circuit $\mathcal{C}_d$:

$$N(d, d') = (t \cdot q)^{\frac{d}{2} - j} \cdot t, \text{ for } d' = 2 \cdot j - 1. \tag{13}$$

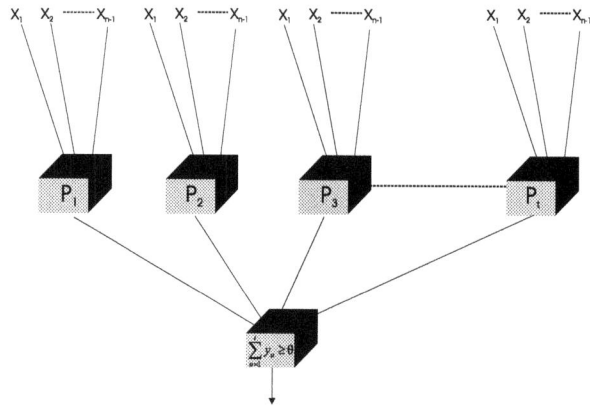

Figure 1. Depth 2 circuit $\mathcal{C}_2$ with $t + 1$ threshold gates.

The Computation of Classification Circuits

The sample set $\mathcal{S}$ is split into two disjoint subsets $\mathcal{S}_L$ and $\mathcal{S}_T$, where $\mathcal{S}_L$ with $|\mathcal{S}_L| \approx \frac{2}{3} \cdot |\mathcal{S}|$ is used to calculate the approximating threshold circuit. The rest of the samples $\mathcal{S}_T$ is used for testing purposes. We assume again that both sets are balanced in the number of positive and negative examples.

In our approach, we try to provide for each perceptron in $\mathcal{C}_d$ its individual training set of approximately the same size p. Therefore, the set $\mathcal{S}_L$ is further partitioned into $d/2$ subsets $\mathcal{S}_L^j$, $1 \leq j \leq d/2$, of approximately equal size $2 \cdot |\mathcal{S}_L| / d$.

From each $\mathcal{S}_L^j$ we calculate $N(d, 2 \cdot j - 1)$ *primary training sets* $T_{[j,i]}^{\text{pr}}$, $1 \leq i \leq N(d, 2 \cdot j - 1)$, by randomly selecting p elements from $\mathcal{S}_L^j$. The parameter p has

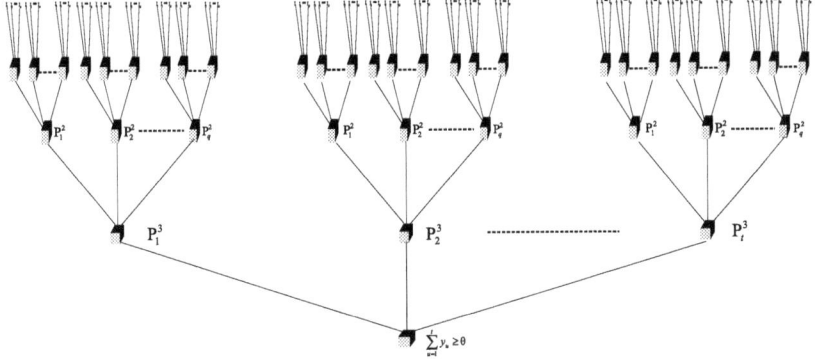

Figure 2. Regular structure of depth 4 circuits C_4.

to be chosen in such a way that

$$p \cdot N(d, 2 \cdot j - 1) \geq \alpha \cdot |S_L^j|, \tag{14}$$

where α is the multiplicity of samples in primary training sets: On average, each sample $\vec{a}$ appears in α different primary training sets. This ensures that a particular $\vec{a}$ is used in different combinations of samples.

We now describe how the actual training sets $T_{[j,i]}^{\text{snd}}$, called *secondary training sets*, are determined from the $T_{[j,i]}^{\text{pr}}$. The $T_{[j,i]}^{\text{snd}}$ are used by the procedure from Section 2.2 as sets S from (2).

1. $j = 1$: In this case we have $T_{[1,i]}^{\text{snd}} = T_{[1,i]}^{\text{pr}}$, $1 \leq i \leq t \cdot (t \cdot q)^{\frac{d}{2}-1}$. The training samples are elements $\vec{a}$ of length $(n-1)$ from S_L^j together with the corresponding classification information $\eta(\vec{a})$, and for each perceptron P_i^1 the procedure from Section 2.2 is applied with respect to $T_{[1,i]}^{\text{pr}}$.

2. $j > 1$: Each perceptron P_i^j, $1 \leq i \leq t \cdot (t \cdot q)^{\frac{d}{2}-j}$, has an associated primary training set $T_{[j,i]}^{\text{pr}}$, and P_i^j defines in C_d q subcircuits of type $C'_{2 \cdot (j-1)}$ with a counting gate as "output". The primary training set $T_{[j,i]}^{\text{pr}}$ is applied to the input nodes of the q subcircuits of type $C'_{2 \cdot (j-1)}$ and therefore produces p tuples of the type $(m_1 m_2 ... m_q)$ with an associated classification $\eta(m_1 m_2 ... m_q)$, which is known from $T_{[j,i]}^{\text{pr}}$. The perceptron P_i^j is trained on $T_{[j,i]}^{\text{snd}} := \{[(m_1 m_2 ... m_q), \eta]\}$ by the procedure from Section 2.2.

Since the q subcircuits of type $C'_{2 \cdot (j-1)}$ are calculated from different training sets at each preceding level, we can expect $m_u \neq m_v$ in $\vec{m} = (m_1 m_2 ... m_q)$, although the components of $\vec{m}$ are calculated from the same sample set.

Parameter Settings

The parameters of the method are Γ, K, t, p, q, α, ϑ, and d, where some of them are only loosely dependent on $\mathcal{S}$.

1. The values of K, q, α, and ϑ are either depending on other parameters or can be chosen relatively independently of the particular sample set, e.g. $q = 3, ..., 7$, $\alpha = 3, ..., 5$, $\vartheta = \lceil t/2 \rceil + 1$. The circuit depth is important for the classification rate, see Section 3, but due to the exponential increase of gates, $d \leq 6$ seems to be a natural limit. Experimental results from [4] suggest $K \leq (|\mathcal{S}|/\delta)^\Gamma$ for the number of transitions in the local search procedure from Section 2.2, where $1 - \delta$ is a confidence parameter, i.e. after K steps the probability to be in an optimum solution is at least $1 - \delta$. Thus, the remaining parameters are Γ, p, and t.

2. The parameter Γ is difficult to estimate a priori. Rough estimations can be derived from the size p of training sets, e.g. $\Gamma = p/3$, which, however, may result in extremely large values for K. On the other hand, one can try to estimate Γ by preliminary experiments on subsets of size p from $\mathcal{S}$: If the current best value $\mathcal{Z}_b$ of the objective function is recorded, one can monitor the maximum increase Δ of the objective function before the next improvement of $\mathcal{Z}_b$ occurs. The maximum value of Δ observed over a sufficently long time period can be used as an upper bound for Γ.

3. The parameter p very much depends on the size of $\mathcal{S}$ and should be as large as possible. The parameters q and t impose a condition on p with respect to the maximum number of secondary training samples $[(m_1 m_2 ... m_q), \eta]$: For m_i we have $-t \leq m_i \leq t$, $1 \leq i \leq q$ and therefore we need to ensure $p \leq (2 \cdot t + 1)^q$, which seems to be valid a priori for the usual range of q and t. In our approach, p is determined by some preliminary experiments: On a depth-two circuit with $t = 50$ we evaluate the average training error on randomly chosen $\tilde{p} := |\mathcal{S}_L|/x$ samples for $x \geq 2$. We take a small x_0 such that the training error stabilises for $x > x_0$. For the two datasets of our study we have $x_0 = 4$ and $x_0 = 6$; see Section 3.3.

4. The parameter t determines the size of the circuit for fixed d and q. The number of couting gates (voting gate as output gate) at depth $d'' = 2 \cdot j$ in circuits $\mathcal{C}_d$ is given by $N(d, 2 \cdot j - 1)/t = (q \cdot t)^{d/2 - j}$; see (13). Thus, the total number of gates in $\mathcal{C}_d$, i.e. the size $S(\mathcal{C}_d)$, is given by

$$S(\mathcal{C}_d) = \sum_{h=0}^{d/2-1} (t+1) \cdot (q \cdot t)^h = (t+1) \cdot \frac{(q \cdot t)^{\frac{d}{2}} - 1}{q \cdot t - 1}. \quad (15)$$

The problem we are specifically interested in is whether circuits $\mathcal{C}_d^t$ with a larger $d > \tilde{d}$ but smaller $t < \tilde{t}$ achieve approximately the same classification rate as $\mathcal{C}_{\tilde{d}}^{\tilde{t}}$ for $S(\mathcal{C}_d^t) << S(\mathcal{C}_{\tilde{d}}^{\tilde{t}})$. Details of experimental results are discussed in Section 3.3.

3 Computational Experiments

For the evaluation of our approach, we chose the splice-junction gene sequences problem from the UCI Machine Learning Repository [30].

3.1 The Splice-junction Gene Sequences Database

The Splice-junction Gene Sequences Database (SJGSD) is taken from

Splice junctions are points on a DNA sequence at which "superfluous" DNA is removed during the process of protein creation in higher organisms. The problem posed in this dataset is to recognize, given a sequence of DNA, the boundaries between exons (the parts of the DNA sequence retained after splicing) and introns (the parts of the DNA sequence that are spliced out). This problem consists of two subtasks: recognizing exon/intron boundaries (referred to as EI sites), and recognizing intron/exon boundaries (IE sites). Additionally, a third class is introduced which is referred to as "Neither". Given a position in the middle of a sequence window, 60 DNA sequence elements are used to decide if this is an IE, EI, or "Neither" class. The database consists of 3190 vectors representing 60 attributes. The class distribution is: 25% for IE (767 instances); 25% for EI (768 instances); and 50% for "Neither" (1655 instances).

In order to discriminate between the three classes, we introduce three databases, each related to a single class as positive examples: the "IE database" consists of 767 positive examples (IE class) and 2423 negative examples (union of EI class and "Neither" class); the "EI database" consists of 768 positive examples and 2422 negative examples; the "Neither database" consists of 1655 positive examples and 1535 negative examples.

Previous results on 25% of samples used as test examples [29] report an error rate on IE, EI and "Neither" classes of 7.55%, 5.74% and 3.99% (best results obtained by different methods), respectively. In [26], a combined correct classification rate of 92.3% is reported.

3.2 Estimations of Circuit Size

Classification problems P as described in Section 3.1 can be encoded as Boolean functions f_P on n input variables where the sample sets usually provide only a tiny fraction of input-output pairs of f_P. For the splice-junction data we have "DNA windows" of length 60 with the usual DNA information from $\{A, C, G, T\}$. Therefore, in binary notation we have $n = 120$.

In both cases the sample data provide only a tiny fraction of the theoretically possible number of function values. A priori, we can argue that not all combinations of binary inputs are feasible (or even a small fraction only has indeed a valid interpretation). We take this into account by simply encoding (enumerating) the sample data instead of using the variable number n as the input to the core of the classificationcircuit: We apply a methodology which is well-known from the synthesis of partially defined Boolean functions [18] in order to obtain some rough estimations of the size of circuits representing the sample data. As before, we denote by $s_L^P := |\mathcal{S}_L|$ the size of the training data set, i.e. we have $s_L^{\text{splice}} = 2127$. By n_P we denote the number of binary variables calculated from the number of input attributes and the range of values each attribute can take, i.e. $n_{\text{splice}} = 120$. The s_L^P training data can be enumerated by using $n_L^P := \lceil \log s_L^P \rceil$ binary variables, i.e. we have $n_L^{\text{splice}} = 12$.

For a given problem P, we introduce the classification circuit $\mathcal{C}_P$: The circuit is built from threshold functions of unbounded fan-in (as basic gates; cf. [19, 21]) and has minimum size with respect to the gate number $\mathcal{S}(\mathcal{C}_P)$. The number of input nodes is n_P. We now try to approximate $\mathcal{C}_P$ by a composition of two circuits

$$\widehat{\mathcal{C}_P} = \mathcal{C}[n_P \to n_L^P] + \mathcal{C}[n_L^P], \tag{16}$$

where $\mathcal{C}[n_P \to n_L^P]$ is a multi-output circuit that calculates the encoding of elements from $\mathcal{S}_L$. The encoding then becomes the binary input to $\mathcal{C}[n_L^P]$. Based on the results about local encoding [18] we assume

$$\mathcal{S}(\mathcal{C}[n_P \to n_L^P]) < \mathcal{S}(\mathcal{C}[n_L^P]) \text{ and therefore } \mathcal{S}(\widehat{\mathcal{C}_P}) < 2 \cdot \mathcal{S}(\mathcal{C}[n_L^P]). \tag{17}$$

Actually, $\mathcal{S}(\mathcal{C}[n_P \to n_L^P])$ depends strongly on the distribution of sample data within the whole domain of feasible samples of P and (17) is valid for sufficiently large n^P and complex P only. Nevertheless, we will focus on $\mathcal{S}(\mathcal{C}[n_L^P])$ alone which seems to be justified by our experimental observation that indeed the chosen problems are associated with complex functions f_P.

To estimate $\mathcal{S}(\mathcal{C}[n_L^P])$, we use the asymptotically optimal design of Boolean functions by liner threshold functions as presented in [19]:

$$\mathcal{S}(f_n) \leq 2 \cdot \sqrt{\frac{2^n}{n}} \cdot \left(1 + \underline{o}(\sqrt{2^n/n})\right), \tag{18}$$

where $f_n = f(x_1, ..., x_n)$ is an arbitrary Boolean function. Moreover, almost all Boolean functions f_n asymptotically require $2 \cdot \sqrt{2^n/n}$ linear threshold gates for their representation [19] (i.e. almost all functions are as complex as the most complex functions).

3.3 Classification Rate vs. Depth

We computed classification results for $\mathcal{C}_2$ and $\mathcal{C}_4$ circuits for a number of different pairs (t, q). According to (13), the number of perceptrons that have to be trained in $\mathcal{C}_4$ is given by $N_{\text{per}} = t \cdot (t \cdot q) + 1$. The value N_{per} was taken as the number of perceptrons in $\mathcal{C}_2$. The remaining parameter settings are $|\mathcal{S}_T| \approx |\mathcal{S}|/3$, $|\mathcal{S}_L^1| = |\mathcal{S}_L^2| = |\mathcal{S}_L|/2$ (for $d = 4$), $K = 20,000$, and $\Gamma = p/3$.

As outlined in Section 2.3, the size p is determined by preliminary experiments on depth-two circuits for $N_{\text{per}} = 50$. The experiments lead us to the following settings: $p_{\text{splice}} := |\mathcal{S}_L|/4$ for all three classes.

Table 1. Error rates on test sets $\mathcal{S}_T^i$, $i = 1, ..., 4$.

N_{per}	Circuits of depth two. Splice-Junction			(t,q)	Circuits of depth four. Splice-Junction		
	IE	EI	Neither		IE	EI	Neither
30	4.1%	3.4%	6.3%	(3,3)	4.3%	3.6 %	7.0%
48	4.0%	3.3%	6.0%	(3,5)	4.2%	3.5 %	6.7%
66	4.0%	3.3%	6.0%	(3,7)	4.1%	3.8 %	7.2%
80	4.0%	3.3%	6.0%	(5,3)	4.1%	3.4 %	7.0%
84	4.0%	3.3%	6.0%	(3,9)	4.5%	3.8 %	7.3%
130	3.9%	3.3%	6.0%	(5,5)	4.0%	3.3 %	6.6%
154	3.9%	3.3%	6.0%	(7,3)	3.9%	3.4 %	6.3%
180	3.8%	3.2%	5.9%	(5,7)	4.0%	3.3 %	6.5%
252	3.8%	3.1%	5.9%	(7,5)	3.7%	3.1 %	6.5%
252	3.8%	3.1%	5.9%	(9,3)	3.7%	3.3 %	6.2%

Let R_d^P denote the maximum of the three best (not necessarily different) classification rates recorded in Table 1 for problem P and circuits $\mathcal{C}_d$. We set for $d = 2, 4$:

$$\mathcal{S}_{\min}^P(d) := \min_{R = R_d^P} \mathcal{S}(\mathcal{C}_d^P[e = R]), \qquad (19)$$

where $e = R$ means an error rate of R by the given circuit. We now allow a margin of deviation from R_d^P by Δ. As an estimation of the circuit size that ensures a high classification rate we take

$$\mathcal{S}_d^P := \max\{\mathcal{S}(\mathcal{C}_d^P[e=R]) : (R_d^P < R \leq R_d^P + \Delta) \& (\mathcal{S}(\mathcal{C}_d^P[e=R]) \leq S_{\min}^P(d))\}. \quad (20)$$

If the max-operation is over an empty set, we take $\mathcal{S}_d^P := \mathcal{S}_{\min}^P(d)$. Taking the max-operation in (20) gives some confidence that the error rates have already stabilised. The calculation of the corresponding values for the two types of circuits and the four classification problems is summarised in Table 2.

Table 2. Circuit size estimates for $\Delta := 0.2\%$.

Circuit size	Splice-Junction		
estimates	IE	EI	Neither
$\mathcal{S}_2^P$	154	154	154
$\mathcal{S}_4^P$	130	84	130

For $P =$ SJGSD we observe an improvement and then stabilisation of error rates with increased circuit size (see Table 1 and Table 2), and we obtain approximately the same size estimations according to (20) for both $d = 2$ and $d = 4$ (except for the EI-class).

We set $\mathcal{S}_P := \max\{\mathcal{S}_2^P, \mathcal{S}_4^P\}$ (see Table 2) and compare the values to $\mathcal{S}_{\text{pred}}^P := x \cdot 2 \cdot (2^{n_L^P}/n_L^P)^{1/2}$; cf. (18), where $x = 4$ is chosen on the following grounds: The RHS of (17) doubles the complexity $\mathcal{S}(\widehat{\mathcal{C}_P})$, and we assume that the third factor on the RHS of (18), which summarises the complexity of auxiliary sub-circuits, at most doubles the product of the first two factors for relatively small values of $(2^{n_L^P}/n_L^P)^{1/2}$. We recall that $n_L^{\text{splice}} = 12$; see Section 3.2. We note that the simple rules from (19) and (20) generate the same $\mathcal{S}_P$ for $P \in \{\text{IE, EI, Neither}\}$; cf. Table 3.

Table 3. Circuit size estimates compared to $\mathcal{S}^P_{\text{pred}}$.

Circuit size estimates	Splice-Junction		
	IE	EI	Neither
$\mathcal{S}_P$	154	154	154
$\mathcal{S}^P_{\text{pred}}$	147	147	147

Thus, we suggest the upper bound

$$\mathcal{S}^P_{\text{pred}} \lessapprox 8 \cdot (2^{n^P_L}/n^P_L)^{1/2} \tag{21}$$

as an *a priori* estimation of the circuit size, where $n^P_L = \lceil \log |\mathcal{S}_L| \rceil$.

In the context of machine learning, the RHS of (18) provides an estimation of the size of elements of the hypotheses space (circuits of threshold functions) that represent particular objects (concepts). The method to prove the lower bound for (18) was utilised in [20, 21] to obtain a lower bound for a sufficient number of samples such that the error rate on test samples is below ε with probability at least $(1-\delta)$. The lower bound is expressed in terms of the VC-dimensions of neural nets [31], which basically equals the number of neurons (threshold gates) [22]. The lower bound is of the type $\max\{4 \cdot n \cdot (n+k)^2/\varepsilon \cdot \log(13/\varepsilon); 4/\varepsilon \cdot \log(2/\delta)\}$, where k is the number of gates in neural nets (which represent the hypotheses). We note that $k = \mathcal{S}^P_{\text{pred}}$ (or even $k = 2 \cdot (2^{n^P_L}/n^P_L)^{1/2}$) and ε in the range of values from Table 1 would result in a number of examples much larger than the number of samples available from our datasets, which is relatively independent of δ.

Finally, we would like to emphasise that our classification results are at least competitive to (or even outperform) methods and results reported in the literature; see Section 3.1.

4 Conclusion

Our computational study on four classification problems from the UCI Machine Learning Repository provides empirical evidence that $8 \cdot (2^{n^P_L}/n^P_L)^{1/2}$ is an appropriate upper bound for the size of threshold circuits in order to achieve a high generalisation capability of circuits. The value of n^P_L is taken as the number of bits necessary to encode all training data by binary strings. The upper bound has been analysed for relatively small n^P_L only. For increasing n^P_L one can expect that the constant becomes significantly smaller (but remains still larger than 2). For the

SJGSD problem we observed that the number of gates in circuits with low error rates is approximately the same for both depth-two and depth-four circuits. The comparison of the circuit size to the asymptotic bound for the most complicated Boolean functions suggests that the chosen problem is indeed of a complex nature.

BIBLIOGRAPHY

[1] E.H.L. Aarts, *Local Search in Combinatorial Optimization* (Wiley & Sons, New York, 1998).

[2] S. Agmon, The relaxation method for linear inequalities, *Canadian J. of Mathem.* 6(1954) 382–392.

[3] A. Albrecht and C.K. Wong, Combining the perceptron algorithm with logarithmic simulated annealing, *Neural Process. Lett.* 14(2001) 75–83.

[4] A. Albrecht and K. Steinhöfel. CT image classification based on logarithmic simulated annealing and Lupanov's threshold circuit theorem, in W.S. Wittig and S. Paul (eds.), *Proc. 8th "Leipziger Informatik-Tage"* (HTWK Leipzig, 2000) pp. 59–65.

[5] E. Allender, The permanent requires large uniform threshold circuits, in *Proc. 2nd Annual International Computing and Combinatorics Conference* (Springer-Verlag, LNCS series, vol. 1090, 1996) pp. 127–135.

[6] E.B. Baum, The perceptron algorithm is fast for nonmalicious distributions, *Neural Computation* 2(1990) 248–260.

[7] A. Blum, A. Frieze, R. Kannan and S. Vempala, A polynomial-time algorithm for learning noisy linear threshold functions, *Algorithmica* 22(1998) 35–52.

[8] A. Blum and R.L. Rivest, Training a 3-node neural network is NP-complete, *Neural Networks* 5(1992) 117–127.

[9] T. Bylander, Learning linear threshold approximations using perceptrons, *Neural Computation* 7(1995) 370–379.

[10] H. Caussinus, P. McKenzie, D. Thérien and H. Vollmer, Nondeterministic NC^1 computation, *J. Comput. System Sci.* 57(1998) 200–212.

[11] V. Černy, A thermodynamical approach to the travelling salesman problem: An efficient simulation algorithm, *J. Optim. Theory Appl.* 45(1985) 41–51.

[12] N. Cristianini and J. Shawe-Taylor, *An Introduction to Support Vector Machines* (Cambridge University Press, 2000).

[13] S.I. Gallant, Perceptron-based learning algorithms, *IEEE Trans. on Neural Networks* 1(1990) 179–191.

[14] B. Hajek, Cooling schedules for optimal annealing, *Mathem. Operat. Res.* 13(1988) 311–329.

[15] K.-U. Höffgen and H.-U. Simon, Robust trainability of single neurons, in *Proc. 5th Annual ACM Workshop on Computational Learning Theory* (1992) pp. 428–439.

[16] K.-U. Höffgen, Computational limitations on training sigmoid neural networks, in *Proc. 2nd Int. Conf. on Artif. Neural Netw.* (1992) pp. 109–112.

[17] S. Kirkpatrick, C.D. Gelatt, Jr. and M.P. Vecchi, Optimization by simulated annealing, *Science* 220(1983) 671–680.

[18] O.B. Lupanov, On a method to design control systems - The local encoding approach (in Russian), *Problemy Kibernetiki* 14(1965) 31–110.

[19] O.B. Lupanov, On the design of circuits by threshold elements (in Russian), *Problemy Kibernetiki* 26(1973) 109–140.

[20] W. Maass, Bounds on the computational power and learning complexity of analog neural nets, in *Proc. 25th Annual ACM Symp. on the Theory of Computing* (1993) pp. 335–344.

[21] W. Maass, On the complexity of learning on neural nets, in *Proc. Computational Learning Theory: EuroColt'93* (Oxford University Press, 1994) pp. 1–17.

[22] W. Maass, R.A. Legenstein and N. Bertschinger, Methods for estimating the computational power and generalization capability of neural microcircuits, in *Proc. Advances in Neural Information Processing Systems* (2005).

[23] M.L. Minsky and S.A. Papert, *Perceptrons* (MIT Press, Cambridge, Massachusetts, 1969).

[24] A. Razborov and A. Wigderson, $n^{\Omega(\log n)}$ lower bounds on the size of depth 3 threshold circuits with AND gates at the bottom, *Inform. Process. Lett.* 45(1993) 303–307.
[25] F. Rosenblatt, *Principles of Neurodynamics* (Spartan Books, New York, 1962).
[26] R. Setiono, Extracting M-of-N rules from trained neural networks, *IEEE Trans. on Neural Networks* 11(2000) 512–519.
[27] J. Shavlik, R.J. Mooney and G. Towell, Symbolic and neural learning programs: An experimental comparison, *Machine Learning* 6(1991) 111–143.
[28] G. Temple, The general theory of relaxations applied to linear systems, *Proc. Roy. Soc. London* 169(1939) 476–500.
[29] G. Towell and J. Shavlik, Knowledge-based artificial neural networks, *Artif. Intell.* 70(1994) 119–165.
[30] UCI Machine Learning Repository: http://www.ics.uci.edu/~mlearn/MLRepository.html
[31] V.N. Vapnik, *Statistical Learning Theory* (Wiley & Sons, New York, 1998).
[32] H. Vollmer, Some Aspects of the Computational Power of Boolean Circuits of Small Depth, Habilitationsschrift, University Würzburg, 1999.
[33] D.H. Wolpert and W.G. Macready, No free lunch theorems for optimization, *IEEE Trans. on Evolutionary Computation* 1(1997) 67–82.

Georgios Lappas
T.E.I. of Western Macedonia
P.O.B. 30, 52100 Kastoria, Greece
Email: lappas@kastoria.teikoz.gr

Kathleen Steinhöfel
King's College London
Dept. of Computer Science
The Strand, London WC2R 2LS, UK
Email: K.Steinhofel@kcl.ac.uk

Andreas A. Albrecht
Univ. of Hertfordshire
School of Computer Science
Hatfield, Herts AL10 9AB, UK
Email: A.Albrecht@herts.ac.uk

Sequence Alignment with Quality Scores

JOONG CHAE NA[1], KANGHO ROH, ALBERTO APOSTOLICO[2], AND KUNSOO PARK

ABSTRACT. In this paper we consider the problem of sequence alignment with quality scores. DNA sequences produced by a base-calling program (as part of sequencing) have quality scores which represent the confidence level for individual bases. However, previous sequence alignment algorithms do not consider such quality scores.

To solve sequence alignment with quality scores, we first consider a more general problem where the input is weighted sequences which are sequences with probabilities that characters occur in each position. We propose a meaningful measure of an alignment of two weighted sequences and show that an optimal alignment in this measure can be found by dynamic programming. Sequence alignment with quality scores can be solved as a special case of the weighted sequence alignment problem.

1 Introduction

Given two sequences A and B, the sequence alignment problem is to find an optimal alignment between them. The alignment of two sequences A and B is a two-row matrix such that the first and second rows contain the characters of A and B in order, respectively, interspersed with some spaces [21]. The score of an alignment is the sum of the scores of its columns. The column score is usually positive for matching letters and negative for distinct letters.

In computational molecular biology, sequence alignment is a fundamental problem since an optimal alignment provides an important measure of similarity between biological sequences [22, 17, 18, 27, 14, 21]. Various notions of alignments and their scoring schemes have been studied (e.g., see [1]). In the early studies, one was concerned about the global alignment of sequences [20]. In many biological applications, however, the local alignment of related sequence

---

[1]Work supported by MOST grant M6-0203-00-0039.
[2]Work Supported in part by an IBM Faculty Partnership Award, by the Italian Ministry of University and Research under the National Projects FIRB RBNE01KNFP, and PRIN "Combinatorial and Algorithmic Methods for Pattern Discovery in Biosequences", and by the Research Program of the University of Padova.

segments is preferred to the global alignment [23, 10]. Another important notion is the multiple alignment that deals with the alignment of multiple sequences [7, 13, 16]. On top of these, there have been vigorous works such as parametric alignments, alignments with gap penalties, normalized alignments, and so on (e.g., see [8, 12, 26, 4, 28, 3, 5]). The alignment problem is closely related to the *longest common subsequences (LCS)* problem in computer science. For the LCS problem, we refer to [24, 22, 2, 9, 15, 25, 19].

We deal with sequence alignment with *quality scores*. In practice, biological data may have errors. For example, DNA sequences produced by a base-calling program (as part of sequencing) have quality scores which represent the confidence pertaining to the individual bases. Most sequence alignment algorithms, however, do not consider such quality scores. Thus these algorithms may not find a best alignment when considering quality scores.

In this paper we present an algorithm that finds an optimal alignment of sequences with quality scores. First, we consider a more general problem where the input is weighted sequences which are sequences with probabilities that characters occur in each position. We propose a meaningful measure of an alignment of two weighted sequences and show that an optimal alignment in this measure can be found by dynamic programming. Then, we apply these results to the original problem, i.e., sequence alignment with quality scores. Although we deal with global alignments in this paper, our algorithms can be easily applied to other kinds of alignments such as local alignments.

The outline of this paper is as follows. In Section 2, we first give some biological background for quality scores and a previous method for sequence alignment. In Section 3, we present a measure of an alignment of weighted sequences and an algorithm to find an optimal alignment. In Section 4, we deal with sequence alignment with quality scores.

2 Preliminaries

We first describe biological circumstances where quality scores are considered and explain the meaning of quality scores. Then, we present a traditional dynamic programming method for global alignments (without quality scores).

2.1 Background

A machine called *sequencer* reads a DNA fragment and produces its *trace* data. The trace data contains the quality of original biological data. Figure 1 shows examples of trace data. As is seen in the figure, the trace data consists of four curves. Each curve represents the signal for one of the four *bases* A, C, T, and G. The horizontal line shows distance and the vertical line shows the level of confidence in the data. The data with higher peaks are more accurate.

Ideally, all peaks in trace data would be spaced an even distance apart and non-

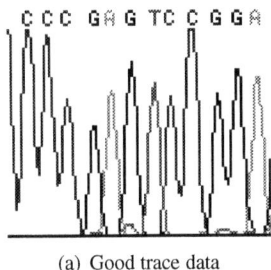 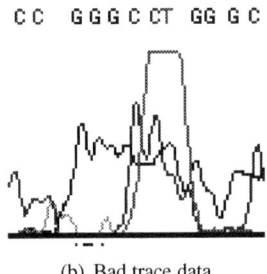

(a) Good trace data   (b) Bad trace data

Figure 1. Examples of trace data.

overlapping. Then we can produce the DNA sequence accurately. Figure 1 (a) shows an example of good trace data. All distances between peaks are almost equivalent and a peak of a curve is much higher than peaks of other curves in each position. But real trace data may deviate from the ideal one because biological experiments can make errors [6]. First, the distances between peaks can be altered, like in Figure 1 (b). For example, the CC segment in the first two positions is much closer than the CG located in the second and third positions. A larger distance may happen due to deletion of a base. Second, there are some positions where two or more curves have similar peaks. Third, there can be positions where the peaks in all four curves are very low.

A program called *base-caller* reads trace data and produces a sequence of bases which represent peaks, like in Figure 1. Most base-callers also produce a quality score corresponding to each base produced, but they don't give quality scores of non-peak bases in each position. Since most base-callers generate similar data, we describe the properties of data of the well-known base-caller *PHRED* [6, 11]. PHRED generates a sequence of bases and a quality score of each base in the sequence. The quality score of PHRED is related to the probability of an error by the formula $Q = -10 \times \log_{10} P$ where P is the probability of an error. For example, a quality score of 20 means that there is a $\frac{1}{100}$ chance that the base has been misidentified. After PHRED generates a sequence and computes quality scores, PHRED executes quality trimming. Toward the end of the trace data, the peaks become progressively less evenly spaced due to errors of biological experiments. Thus each end of a sequence usually has noisy data which have very low quality scores. Since noisy data can have bad influences, PHRED cuts off each end of a sequence by quality trimming. So each base of a resulting sequence has a meaningfully high score. Quality scores of most bases after quality trimming are larger than 10.

2.2 Global Alignment

We first give some definitions and notations. Let the alphabet Σ be $\{a, c, t, g\}$. A *sequence* is concatenations of zero or more characters from an alphabet Σ. A space is denoted by Δ. The length of a sequence A is denoted by $|A|$. We denote ith character of a sequence A by A_i and a subsequence $A_i A_{i+1} \cdots A_j$ by $A[i..j]$.

Given two sequences $A = A_1 A_2 \cdots A_m$ and $B = B_1 B_2 \cdots B_n$ ($|A| = m$, $|B| = n$), an *alignment* of A and B is denoted by $A^* = A_1^* A_2^* \cdots A_l^*$ and $B^* = B_1^* B_2^* \cdots B_l^*$ ($n, m \leq l$). It is constructed by inserting zero or more *spaces* (Δ) into A and B so that A_i^* maps to B_i^* for $1 \leq i \leq l$. There are three kinds of mappings for A_i^* and B_i^*.

- match : $A_i^* = B_i^* \neq \Delta$.
- mismatch : $A_i^* \neq \Delta$, $B_i^* \neq \Delta$ and $A_i^* \neq B_i^*$.
- indel : ($A_i^* = \Delta$ and $B_i^* \neq \Delta$) or ($A_i^* \neq \Delta$ and $B_i^* = \Delta$).

Note that we do not allow the case of $A_i^* = B_i^* = \Delta$. Each type of mapping has a mapping score: a positive match score γ, a negative mismatch score δ and a negative gap score μ.

We define the *alignment score* $S(A^*, B^*)$ of A^* and B^* as the sum of mapping scores of all positions. That is,

$$S(A^*, B^*) = \sum_{i=1}^{l} S(A_i^*, B_i^*).$$

EXAMPLE 4.1. Let A be aat and B be $acat$. The following is an alignment of A and B.

$$
\begin{array}{ccccc}
A = & a & a & \Delta & t \\
  & | & | & | & | \\
B = & a & c & a & t
\end{array}
$$

The first and the fourth are match mappings, the second is a mismatch mapping, and the third is an indel mapping. Then, given $\gamma = 1$, $\delta = -1$ and $\mu = -2$, the score of this alignment is $1 - 1 - 2 + 1 = -1$.

Given two sequences A and B, and score parameters γ, δ, μ, the sequence alignment problem is to find an alignment of maximum score among all possible alignments of A and B. We can find an optimal alignment of sequences by dynamic programming. Let $S(A, B)$ denote the maximum score for sequences A and B. We computes $S(A, B)$ by dynamic programming filling a table of size $(m + 1) \times (n + 1)$. Specifically, let $H_{i,j}$ denote $S(A[1..i], B[1..j])$ for

Algorithm 1 Recurrence relation for sequence alignment

$\gamma > 0$: match score.
$\delta < 0$: mismatch score.
$\mu < 0$: gap score.

$$S(A_i, B_j) = \begin{cases} \gamma & \text{if } A_i \text{ and } B_j \text{ are a match} \\ \delta & \text{if } A_i \text{ and } B_j \text{ are a mismatch} \end{cases}.$$

$H_{0,0} = 0$.
$H_{i,0} = H_{i-1,0} + \mu, (1 \leq i \leq m)$.
$H_{0,j} = H_{0,j-1} + \mu, (1 \leq j \leq n)$.

$$H_{i,j} = \max \begin{cases} H_{i-1,j-1} + S(A_i, B_j) \\ H_{i-1,j} + \mu \\ H_{i,j-1} + \mu \end{cases} \quad (1 \leq i \leq m, 1 \leq j \leq n).$$

$0 \leq i \leq m, 0 \leq j \leq n$. $H_{i,j}$ can be computed by Algorithm 1. An optimal alignment can be found by tracing the table from $H_{m,n}$ to $H_{0,0}$. This algorithm takes $O(mn)$ time because it computes a table of size $O(mn)$ [20].

3 Alignment of Weighted Sequences

We formally define a weighted sequence as follows. A *weighted sequence* $A = A_1 A_2 \cdots A_n$ is a nondeterministic sequence in which each A_i has a probability set $\{P_a(A_i), P_c(A_i), P_t(A_i), P_g(A_i), P_-(A_i)\}$ such that the sum of the five probabilities is 1, where $P_x(A_i)$ is the probability that a solid character $x \in \Sigma$ occurs in the ith position and $P_-(A_i)$ is the probability that no solid characters occur in the ith position. The space originated from no occurrence of solid characters is denoted by '−' to distinguish it from the space (Δ) inserted in an alignment in Section 2.2. We call A_i a *weighted character*.

EXAMPLE 4.2. Let A be $A_1 A_2 A_3$. Assume that the probability of each A_i is as follows.

	A_1	A_2	A_3
a	1	1	0
c	0	0	0.3
t	0	0	0.5
g	0	0	0
–	0	0	0.2

Then, A can be one of aaa, aat, and aa with probabilities 0.3, 0.5, and 0.2, respectively.

An *alignment* of two weighted sequences $A = A_1 A_2 \cdots A_m$ and $B = B_1 B_2 \cdots B_n$ ($|A| = m$, $|B| = n$) is denoted by $A^* = A_1^* A_2^* \cdots A_l^*$ and $B^* = B_1^* B_2^* \cdots B_l^*$ ($n, m \leq l$). It is constructed by inserting zero or more spaces (Δ) into A and B so that A_i^* maps to B_i^* for $1 \leq i \leq l$.

In any alignment, $A_i^* = B_i^* = \Delta$ is not allowed, whence there are only two kinds of mappings for A_i^* and B_i^*.

- regular: $A_i^* \neq \Delta$ and $B_i^* \neq \Delta$.
- indel: ($A_i^* = \Delta$ and $B_i^* \neq \Delta$) or ($A_i^* \neq \Delta$ and $B_i^* = \Delta$).

A mapping of A_i^* and B_i^* may take up one of several *submappings*, depending on which pair from $\Sigma \cup \{-\}$ gives values to A_i^* and B_i^*. Specifically, we distinguish the following cases when a_i^* and b_i^* are the character specifications for A_i^* and B_i^*, respectively.

- match: $a_i^* = b_i^* \neq \text{'--'}$.
- mismatch: $a_i^* \neq \text{'--'}$, $b_i^* \neq \text{'--'}$ and $a_i^* \neq b_i^*$.
- gap: ($a_i^* = \text{'--'}$ and $b_i^* \neq \text{'--'}$) or ($a_i^* \neq \text{'--'}$ and $b_i^* = \text{'--'}$).
- empty: $a_i^* = b_i^* = \text{'--'}$.

Note that, with the exception of the empty one, the above submappings are the same as those found in a usual alignment of sequences that are composed of solid characters in Section 2.2. Likewise, each submapping has a score: the match score γ, the mismatch score δ and the gap score μ. The score of the empty submapping is 0 because this can be discarded in an alignment.

We define $S(A_i^*, B_i^*)$ of A_i^* and B_i^* as the expected submapping score. Specifically, let P_m, P_n and P_g be the probabilities that a match submapping, a mismatch

Table 1. Probabilities of a mapping of two weighted characters.

	a : α_a	c : α_c	t : α_t	g : α_g	-: α_-
a : β_a	$\alpha_a\beta_a$	$\alpha_c\beta_a$	$\alpha_t\beta_a$	$\alpha_g\beta_a$	$\alpha_-\beta_a$
c : β_c	$\alpha_a\beta_c$	$\alpha_c\beta_c$	$\alpha_t\beta_c$	$\alpha_g\beta_c$	$\alpha_-\beta_c$
t : β_t	$\alpha_a\beta_t$	$\alpha_c\beta_t$	$\alpha_t\beta_t$	$\alpha_g\beta_t$	$\alpha_-\beta_t$
g : β_g	$\alpha_a\beta_g$	$\alpha_c\beta_g$	$\alpha_t\beta_g$	$\alpha_g\beta_g$	$\alpha_-\beta_g$
- : β_-	$\alpha_a\beta_-$	$\alpha_c\beta_-$	$\alpha_t\beta_-$	$\alpha_g\beta_-$	$\alpha_-\beta_-$

submapping and a gap submapping occur, respectively. If the mapping of A_i^* and B_i^* is regular, then

$$S(A_i^*, B_i^*) = \gamma P_m + \delta P_n + \mu P_g;$$

otherwise,

$$S(A_i^*, B_i^*) = \mu P_g.$$

We compute the probabilities of a match, a mismatch and a gap in a regular mapping of two weighted characters α and β (i.e., $\alpha = A_i^*$, $\beta = B_i^*$). Table 1 shows the probabilities of all possible combinations of α and β. For notational convenience, we denote $P_x(\alpha)$ by α_x for $x \in \Sigma \cup \{-\}$ in Table 1. From Table 1, we can get $P_m(\alpha, \beta), P_n(\alpha, \beta)$, and $P_g(\alpha, \beta)$ as follows:

- $P_m(\alpha, \beta) = \alpha_a\beta_a + \alpha_c\beta_c + \alpha_t\beta_t + \alpha_g\beta_g$

- $P_n(\alpha, \beta) = \alpha_a\beta_c + \alpha_a\beta_t + \alpha_a\beta_g$
  $+ \alpha_c\beta_a + \alpha_c\beta_t + \alpha_c\beta_g$
  $+ \alpha_t\beta_a + \alpha_t\beta_c + \alpha_t\beta_g$
  $+ \alpha_g\beta_a + \alpha_g\beta_c + \alpha_g\beta_t$

- $P_g(\alpha, \beta) = \alpha_-\beta_a + \alpha_-\beta_c + \alpha_-\beta_t + \alpha_-\beta_g$
  $+ \alpha_a\beta_- + \alpha_c\beta_- + \alpha_t\beta_- + \alpha_g\beta_-$
  $= \alpha_-(1 - \beta_-) + \beta_-(1 - \alpha_-)$

In case of an indel mapping, we resort to the following expression. Without loss of generality, we assume that β is Δ. Then, $P_m(\alpha, \Delta)$ and $P_n(\alpha, \Delta)$ is 0, and $P_g(\alpha, \Delta)$ is $1 - \alpha_-$ because β_- is 1. Thus, $S(\alpha, \Delta)$ is $\mu \times (1 - \alpha_-)$.

We define the *alignment score* $S(A^*, B^*)$ of A^* and B^* as the sum of mapping scores of all positions. That is,

$$S(A^*, B^*) = \sum_{i=1}^{l} S(A_i^*, B_i^*).$$

EXAMPLE 4.3. Let A^* be $A_1 A_2 \Delta A_3$ and B^* be $B_1 B_2 B_3 B_4$. Suppose that the probability set is as follows.

	A_1	A_2	A_3	B_1	B_2	B_3	B_4
a	1	1	0.5	1	0	1	0
c	0	0	0	0	0.6	0	0
t	0	0	0.5	0	0	0	1
g	0	0	0	0	0	0	0
−	0	0	0	0	0.4	0	0

Then, given $\gamma = 1, \delta = -1$ and $\mu = -2$, the mapping score of each position is as follows:

$S(A_1^*, B_1^*) = S(A_1, B_1) = 1,$
$S(A_2^*, B_2^*) = S(A_2, B_2) = -1 \times 0.6 - 2 \times 0.4 = -1.4,$
$S(A_3^*, B_3^*) = S(\Delta, B_3) = -2,$ and
$S(A_4^*, B_4^*) = S(A_3, B_4) = 1 \times 0.5 - 1 \times 0.5 = 0.$

The score of this alignment is $1 - 1.4 - 2 + 0 = -2.4$.

Given two weighted sequences A and B, and score parameters γ, δ, μ, the weighted sequence alignment problem is to find an alignment of maximum score among all possible alignments of A and B. An optimal alignment of weighted sequences and its score can be found by Algorithm 2. It is the same as Algorithm 1 except for computing mapping scores. Hence, Algorithm 2 takes $O(mn)$ time and space.

4 Sequence Alignment with Quality Scores

In this section we introduce the sequence alignment problem with quality scores and convert it into the sequence alignment problem of weighted sequences. Then we present an algorithm for solving sequence alignment with quality scores.

A sequence $A = A_1 A_2 \cdots A_n$ with quality scores is a sequence in which each A_i is a solid character in Σ and has a quality score Q_{A_i}. The score Q_{A_i} means that the probability of error associated with A_i is $10^{\frac{-Q_{A_i}}{10}}$.

Algorithm 2 Recurrence relation for the weighted sequence alignment problem

$\gamma > 0$: match score.
$\delta < 0$: mismatch score.
$\mu < 0$: gap score.

P_m : match probability.
P_n : mismatch probability.
P_g : gap probability.

Δ : a special character whose gap probability is 1.

$S(A_i, B_j) = \gamma \times P_m(A_i, B_j) + \delta \times P_n(A_i, B_j) + \mu \times P_g(A_i, B_j)$: score of regular mapping.
$S(A_i, \Delta) = \mu \times P_g(A_i, \Delta)$: score of indel mapping.

$H_{0,0} = 0$.
$H_{i,0} = H_{i-1,0} + S(A_i, \Delta), (1 \leq i \leq m)$.
$H_{0,j} = H_{0,j-1} + S(B_j, \Delta), (1 \leq j \leq n)$.

$$H_{i,j} = \max \begin{cases} H_{i-1,j-1} + S(A_i, B_j) \\ H_{i-1,j} + S(A_i, \Delta) \\ H_{i,j-1} + S(B_j, \Delta) \end{cases} \quad (1 \leq i \leq m, 1 \leq j \leq n).$$

EXAMPLE 4.4. Let $A = atg$ with quality scores 10, 20, and 30 for a, t and g, respectively. Then the error probability of each character is as follows.

sequence	a	t	g
quality score	10	20	30
error probability	0.1	0.01	0.001

A character $x \in \Sigma$ with quality score Q_x can be changed into a weighted character α with a probability set $\{P_a(\alpha), P_c(\alpha), P_t(\alpha), P_g(\alpha), P_-(\alpha)\}$, as follows. A character $x \in \Sigma$ in a position of a given sequence A will be called the *representative* for that position. Suppose that representative x in a position is base a. By definition of quality score, $P_a(\alpha)$ is $1 - 10^{\frac{-Q_x}{10}}$, but we do not know the remaining probabilities $\{P_c(\alpha), P_t(\alpha), P_g(\alpha), P_-(\alpha)\}$. We make the following assumptions in order to determine these probabilities.

- Assumption 1: The probability of a representative in a position is much higher than those of other characters (i.e, other elements in $\Sigma \cup \{-\}$) in that position. The quality score is usually larger than 10. Hence the probability of a representative is usually higher than 0.9.

- Assumption 2: The characters in $\Sigma \cup \{-\}$ except a representative have the same probability. Because the probability of a representative is large, the probabilities of other characters in $\Sigma \cup \{-\}$ are very small, and thus their differences are negligible.

From Assumption 2, we get $P_c(\alpha) = P_t(\alpha) = P_g(\alpha) = P_-(\alpha) = \frac{10^{-\frac{Q_x}{10}}}{4}$. By transforming every character with a quality score into a weighted character, we get a weighted sequence.

EXAMPLE 4.5. Let $A = atg$ with quality scores 10, 20, and 30, for a, t and g, respectively. We convert A to a weighted sequence $A_1 A_2 A_3$. The following table shows the probability sets of the weighted sequence.

sequence	a	t	g
quality	10	20	30
weighted sequence	A_1	A_2	A_3
$P_a(A_i)$	0.9	0.0025	0.00025
$P_c(A_i)$	0.025	0.0025	0.00025
$P_t(A_i)$	0.025	0.99	0.00025
$P_g(A_i)$	0.025	0.0025	0.999
$P_-(A_i)$	0.025	0.0025	0.0009

The sequence alignment problem with quality scores is to find an optimal alignment of two sequences that have quality scores. To solve this problem, we convert the two input sequences into weighted sequences. Then we compute an optimal alignment by Algorithm 2 of Section 3.

Let $A = A_1 A_2 \cdots A_m$ and $B = B_1 B_2 B_3 \cdots B_n$ be (converted) weighted sequences. Let $A^* = A_1^* A_2^* \cdots A_l^*$ and $B^* = B_1^* B_2^* \cdots B_l^*$ ($1 \leq m, n \leq l$) be an alignment of A and B. Since representatives are given, we can divide a regular mapping of A_i^* and B_i^* into two cases: a regular-match and a regular-mismatch. A regular-match is the case that the representatives are the same and a regular-mismatch is the case that the representatives are different. So there are three kinds of mappings of A_i^* and B_i^*. Note that $A_i^* = B_i^* = \Delta$ is not allowed.

- regular-match: $A_i^* \neq \Delta$ and $B_i^* \neq \Delta$, and the representatives of A_i^* and B_i^* are the same.
- regular-mismatch: $A_i^* \neq \Delta$ and $B_i^* \neq \Delta$, and the representatives of A_i^* and B_i^* are different.
- indel : ($A_i^* = \Delta$ and $B_i^* \neq \Delta$) or ($A_i^* \neq \Delta$ and $B_i^* = \Delta$).

Now we compute the score of each kind of mapping. Let α and β be weighted characters A_i^* and B_i^*, respectively. Again, we denote $P_\alpha(x)$ by α_x.

- Regular-match case: Suppose that the representatives of α and β are a. Then we get $\alpha_c = \alpha_t = \alpha_g = \alpha_- = \frac{1-\alpha_a}{4}$ and $\beta_c = \beta_t = \beta_g = \beta_- = \frac{1-\beta_a}{4}$. The probability table is as follows.

	a	c	t	g	-
a	$\alpha_a \beta_a$	X	X	X	X
c	Y	Z	Z	Z	Z
t	Y	Z	Z	Z	Z
g	Y	Z	Z	Z	Z
-	Y	Z	Z	Z	Z

$X = \frac{(1-\alpha_a)\beta_a}{4}, Y = \frac{\alpha_a(1-\beta_a)}{4}, Z = \frac{(1-\alpha_a)(1-\beta_a)}{16}$.

Since $Z = \frac{(1-\alpha_a)(1-\beta_a)}{16}$ is much smaller than other terms, we ignore that in the following computations. From the above table we get:

1. $P_m(\alpha, \beta) = \alpha_a \beta_a + 3Z \approx \alpha_a \beta_a$

2. $P_n(\alpha, \beta) = 3X + 3Y + 6Z \approx \left(\frac{\alpha_a + \beta_a - 2\alpha_a \beta_a}{4}\right) \times 3$

3. $P_g(\alpha, \beta) = X + Y + 6Z \approx \frac{\alpha_a + \beta_a - 2\alpha_a \beta_a}{4}$.

Therefore, the mapping score of a regular-match case is

$$S(\alpha, \beta) = \gamma \times \alpha_a \beta_a + \delta \times \frac{3(\alpha_a + \beta_a - 2\alpha_a \beta_a)}{4}$$
$$+ \mu \times \frac{\alpha_a + \beta_a - 2\alpha_a \beta_a}{4}.$$

- Regular-mismatch case: Suppose that the representatives of α and β are c and a, respectively. We get $\alpha_a = \alpha_t = \alpha_g = \alpha_- = \frac{1-\alpha_c}{4}$ and $\beta_c = \beta_t = \beta_g = \beta_- = \frac{1-\beta_a}{4}$. The probability table is as follows.

	a	c	t	g	-
a	X	$\alpha_c\beta_a$	X	X	X
c	Z	Y	Z	Z	Z
t	Z	Y	Z	Z	Z
g	Z	Y	Z	Z	Z
-	Z	Y	Z	Z	Z

$$X = \frac{(1-\alpha_c)\beta_a}{4}, Y = \frac{\alpha_c(1-\beta_a)}{4}, Z = \frac{(1-\alpha_c)(1-\beta_a)}{16}.$$

From the table we get:

1. $P_m(\alpha, \beta) = X + Y + 2Z \approx \frac{\alpha_c+\beta_a-2\alpha_c\beta_a}{4}$
2. $P_n(\alpha, \beta) = \alpha_c\beta_a + 2X + 2Y + 7Z \approx \frac{\alpha_c+\beta_a}{2}$
3. $P_g(\alpha, \beta) = X + Y + 6Z \approx \frac{\alpha_c+\beta_a-2\alpha_c\beta_a}{4}$

The mapping score of a regular-mismatch case is

$$S(\alpha, \beta) = \gamma \times \frac{\alpha_c + \beta_a - 2\alpha_c\beta_a}{4} + \delta \times \frac{\alpha_c + \beta_a}{2}$$
$$+ \mu \times \frac{\alpha_c + \beta_a - 2\alpha_c\beta_a}{4}.$$

- An indel case: Suppose that $\beta = \Delta$ and the representative of α is a. Then,

$$S(\alpha, \beta) = \mu \times (1 - \alpha_-) = \mu \times \frac{3 + \alpha_a}{4}.$$

We further simplify the mapping scores of regular-match and regular-mismatch cases. Consider a regular-match case. The values α_a and β_a are usually larger than 0.9. When α_a and β_a are 0.9, $P_m(\alpha, \beta) \approx \alpha_a\beta_a = 0.81$ and $P_g(\alpha, \beta) \approx \frac{1}{4}(\alpha_a + \beta_a - 2\alpha_a\beta_a) = 0.045$ and $P_n(\alpha, \beta) \approx \frac{3}{4}(\alpha_a + \beta_a - 2\alpha_a\beta_a) = 0.135$. Since $P_n(\alpha, \beta)$ and $P_g(\alpha, \beta)$ are much smaller than $P_m(\alpha, \beta)$, we simplify the mapping score $S(\alpha, \beta)$ to $\gamma P_m(\alpha, \beta) = \gamma \times \alpha_a\beta_a$. In a regular-mismatch case, we also ignore the $\frac{1}{4}(\alpha_c + \beta_a - 2\alpha_c\beta_a)$ term and simplify the mapping score $S(\alpha, \beta)$ to $\delta \times P_n(\alpha, \beta) = \delta \times \frac{\alpha_c + \beta_a}{2}$.

EXAMPLE 4.6. Let A be aat and B be $acat$. Suppose that the quality scores of the characters of A are 10, 20, and 10 in order, and those of B are 10, 10, 20, and 10. The following is an alignment of A and B.

$$A = a \quad a \quad \Delta \quad t$$
$$| \quad | \quad | \quad |$$
$$B = a \quad c \quad a \quad t$$

The first and the fourth are regular-match mappings, the second is a regular-mismatch mapping, and the third is an indel mapping. Then, given $\gamma = 1$, $\delta = -1$ and $\mu = -2$, the mapping score of each position is as follows:

$S(A_1^*, B_1^*) = 1 \times (0.9 \times 0.9) = 0.81$,
$S(A_2^*, B_2^*) = -1 \times \frac{(0.99 + 0.9)}{2} = -0.945$,
$S(A_3^*, B_3^*) = -2 \times \frac{(3 + 0.99)}{4} = -1.995$, and
$S(A_4^*, B_4^*) = 1 \times (0.9 \times 0.9) = 0.81$.

The score of this alignment is $0.81 - 0.945 - 1.995 + 0.81 = -1.32$.

Algorithm 3 shows the recurrence relation for sequence alignment with quality scores after these simplifications. Notice that Algorithm 1 is a special case of Algorithm 3. When P_{A_i}'s and P_{B_j}'s are 1, the recurrence relation of Algorithm 3 is the same as that of Algorithm 1. From the viewpoint of Algorithm 3, Algorithm 1 solves the alignment problem on the assumption that the error probabilities of representatives are 0.

5 Experimental Results

In this section we show experimental results. The objective of our experiments is to check the following two questions. One is how often the case occurs that an optimal alignment by Algorithm 1 is different from that by Algorithm 3. The other question is which algorithm finds better alignments.

We performed experiments in the following way. We obtained four real genome data from the NCBI website (http://www.ncbi.nlm.nih.gov/Traces/trace.fcgi). The names of the data are "gnl|ti|14264831" (Data Set 1), "gnl|ti|222680126" (Data Set 2), "gnl|ti|312912383" (Data Set 3), and "gnl|ti|125121088" (Data Set 4). Each data set consists of 20000 sequences with quality scores. The average length of sequences in each data set is about 1400 ~ 2000. The averages of quality scores in Data Set1, Data Set2, Data Set3, and Data Set4 are 11.9, 17.7, 16.3, and 18.0, respectively.

Algorithm 3 Recurrence relation for sequence alignment with quality scores

$\gamma > 0$: match score.
$\delta < 0$: mismatch score.
$\mu < 0$: gap score.

Q_x : the quality score of a character x.
$P_x = 1 - 10^{\frac{-Q_x}{10}}$.

$$S(A_i, B_j) = \begin{cases} \gamma \times P_{A_i} P_{B_j} & \text{if } A_i \text{ and } B_j \text{ are a match} \\ \delta \times \frac{P_{A_i} + P_{B_j}}{2} & \text{if } A_i \text{ and } B_j \text{ are a mismatch} \end{cases}.$$

$H_{0,0} = 0.$
$H_{i,0} = H_{i-1,0} + (\mu \times \frac{3 + P_{A_i}}{4}), (1 \leq i \leq m).$
$H_{0,j} = H_{0,j-1} + (\mu \times \frac{3 + P_{B_j}}{4}), (1 \leq j \leq n).$

$$H_{i,j} = \max \begin{cases} H_{i-1,j-1} + S(A_i, B_j) \\ H_{i-1,j} + (\mu \times \frac{3 + P_{A_i}}{4}) \\ H_{i,j-1} + (\mu \times \frac{3 + P_{B_j}}{4}) \end{cases} \quad (1 \leq i \leq m, 1 \leq j \leq n).$$

We could observe that the outputs of Algorithm 3 were totally different from those of Algorithm 1. We selected randomly 2000 pairs of sequences in each data set, and computed and compared optimal alignments of these pairs by Algorithm 1 and Algorithm 3. A pair of sequences is one of the following two types.

- S-type: An optimal alignment by Algorithm 1 is the same as that by Algorithm 3.

- D-type: An optimal alignment by Algorithm 1 is different from that by Algorithm 3.

To our surprise, all 8000 pairs were D-type in the above experiments. We also experimented on sequences of short length, which were made by cutting the sequences of the genome data. Table 2 shows the results. Even on pairs of sequences of length 10, about 50% pairs were D-type. The time taken by Algorithm 3 was slightly larger than that by Algorithm 1, but the difference was negligible.

In case of D-type pairs, we could verify that the alignments by Algorithm 3 are better than those by Algorithm 1 when considering quality scores. We present two examples, which are quite typical in short sequences. The first example is shown in Figure 2. Alignments U and V are the optimal alignments computed by Algorithm

Table 2. The number of D-type and S-type pairs of short sequences.

Data Set	length: 10		length: 20		length: 30	
	S-type	D-type	S-type	D-type	S-type	D-type
1	803	1197	205	1795	81	1919
2	1319	681	629	1371	285	1715
3	805	1195	219	1781	82	1918
4	1143	857	497	1503	143	1857

1 and Algorithm 3, respectively. The numerals represent the quality score of each character. In (regular) match and (regular) mismatch mappings, the pair of quality scores in a mapping will be called a *quality-score-pair* of the mapping and denoted by (x, y), where x is the score of one character and y is the score of the other. For example, the quality-score-pair of the first mapping in Alignment U is (35,25). The box indicates the parts of U and V where mappings are different from each other. The boxes of alignments U and V equally consist of one match mapping and three indel mappings. However, the quality-score-pair (34,56) of the match mapping in alignment U is worse than that (35,56) in alignment V. Also, the quality scores 35, 34, 34 of the indel mappings in alignment U are higher than 34, 34, 34 in alignment V. Therefore, it is obvious that alignment V is better than alignment U when quality scores are considered.

Figure 3 shows a more complicated example. Alignments X and Y are the optimal alignments computed by algorithm 1 and algorithm 3, respectively. The boxes of alignments X and Y equally consist of three match, three mismatch, and two indel mappings. Consider the quality scores of mappings in the boxes. In case of match mappings, the quality-score-pairs of X are (46,13), (46,19), and (46,29), and those of Y are (46,13), (46,19), and (46,39). Thus, alignment Y is better than alignment X in the part of match mappings. In case of mismatch mappings, the quality-score-pairs of X are (46,8) and (46,39), and those of Y are (39,8) and (46,29). Hence, alignment Y is better than alignment X in mismatch mappings. In case of indel mappings, the quality scores of X are 39 and 46, and those of Y are 46 and 46. Although alignment Y is worse than alignment X in indel mappings, the superiority of X in this case is weaker than that of alignment Y in match and mismatch mappings. Therefore, alignment Y is better than alignment X when we consider all mappings.

Alignment U

		35	39	39	35	34	34	34
$A =$		t	t	t	c	a	c	t
$B =$		c	t	t	Δ	Δ	c	Δ
		25	39	39			56	

Alignment V

		35	39	39	35	34	34	34
$A =$		t	t	t	c	a	c	t
$B =$		c	t	t	c	Δ	Δ	Δ
		25	39	39	56			

Figure 2. Given $\gamma = 1$, $\delta = -1$ and $\mu = -1$, U and V are the alignments computed by algorithm 1 and algorithm 3, respectively.

Alignment X

	39	39	39	46	46	46	46	46	46	46	40
$C =$	c	t	t	c	c	c	t	a	g	t	t
$D =$	c	Δ	Δ	a	c	c	Δ	a	a	a	a
	6			8	13	19		29	39	39	33

Alignment Y

	39	39	39	46	46	46	46	46	46	46	40
$C =$	c	t	t	c	c	c	t	a	g	t	t
$D =$	c	Δ	a	c	c	a	a	a	Δ	Δ	a
	6		8	13	19	29	39	39			33

Figure 3. Given $\gamma = 1$, $\delta = -1$ and $\mu = -1$, X and Y are the alignments computed by algorithm 1 and algorithm 3, respectively.

BIBLIOGRAPHY

[1] A. Apostolico and R. Giancarlo. Sequence alignment in molecular biology. *Journal of Computational Biology*, 5(2):173–196, 1998.

[2] A. Apostolico and C. Guerra. The longest common subsequence problem revisited. *Algorithmica*, 2:315–336, 1987.

[3] A. N. Arslan, Ö. Eğecioğlu, and P. A. Pevzner. A new approach to sequence comparison: Normalized sequence alignment. *Bioinformatics*, 17(4):327–337, 2001.

[4] G. Benson. Sequence alignment with tandem duplication. In *Proceedings of the First Annual International Conference on Computational Molecular Biology (RECOMB-97)*, pages 27–36, 1997.

[5] M. Crochemore, G. M. Landau, and M. Ziv-Ukelson. A sub-quadratic sequence alignment algorithm for unrestricted cost matrices. In *Proceedings of the 13th ACM-SIAM Symposium on Discrete Algorithms*, pages 679–688, 2002.

[6] B. Ewing, L. Hillier, M. C. Wendl, and P. Green. Base-calling of automated sequencer traces using *phred*. I. accuracy assessment. *Genome Research*, 8(3):175–185, 1998.

[7] D. Feng and R. Doolittle. Progressive sequence alignment as a prerequisite to correct phylogenetic trees. *Journal of Molecular Evolution*, 25:351–360, 1987.

[8] W. M. Fitch and T. F. Smith. Optimal sequence alignments. *Proceedings of the National Academy of Sciences USA*, 80:1382–1386, 1983.

[9] Z. Galil and R. Giancarlo. Data structures and algorithms for approximate string matching. *Journal of Complexity*, 4(1):33–72, 1988.

[10] O. Gotoh. An improved algorithm for matching biological sequences. *Journal of Molecular Biology*, 162(3):705–708, 1982.

[11] P. Green. Documentation for phrap. Genome Center, University of Washington, 1995. http://www.phrap.org/phrap.docs/phrap.html.

[12] D. Gusfield. Parametric combinatorial computing and a problem of program module distribution. *Journal of the ACM*, 30(3):551–563, 1983.

[13] D. Gusfield. Efficient methods for multiple sequence alignment with guaranteed error bounds. *Bulletin of Mathematical Biology*, 55(1):141–154, 1993.

[14] D. Gusfield. *Algorithms on Strings, Tree, and Sequences*. Cambridge University Press, Cambridge, 1997.

[15] D. S. Hirschberg and L. L. Larmore. The set-set LCS problem. *Algorithmica*, 4(4):503–510, 1989.

[16] T. J. Hubbard, A. M. Lesk, and A. Tramontano. Gathering them into the fold. *Nature Structural Biology*, 4:313, 1996.

[17] G. Myers. An overview of sequence comparison algorithms in molecular biology. Technical Report TR-91-29, Dept. of Computer Science, University of Arizona, 1991.

[18] G. Myers. *Computational Methods in Genome Research*, chapter Algorithmic Advances for Searching Biosequence Databases, pages 121–135. Plenum Press, New York, 1994.

[19] G. Navarro. A guided tour to approximate string matching. *ACM Computing Surveys*, 33(1):31–88, 2001.

[20] S. B. Needleman and C. D. Wunsch. A general method applicable to the search for similarities in the amino acid sequences of two proteins. *Journal of Molecular Biology*, 48(3):443–453, 1970.

[21] P. A. Pevzner. *Computational Molecular Biology: An Algorithmic Approach*. The MIT Press, Cambridge, 2000.

[22] D. Sankoff and J. Kruskal. *Time Warps, String Edits, and Macromolecules: The Theory and Practice of Sequence Comparison*. Addison-Wesley, Reading, MA, 1983.

[23] T. F. Smith and M. S. Waterman. Identification of common molecular subsequences. *Journal of Molecular Biology*, 147(1):195–197, 1981.

[24] R. Wagner and M. Fisher. The string to string correction problem. *Journal of the ACM*, 21(1):168–173, 1974.

[25] B. F. Wang, G. H. Chen, and K. Park. On the set LCS and set-set LCS problems. *Journal of Algorithms*, 14(3):466–477, 1993.

[26] M. S. Waterman. Efficient sequence alignment algorithms. *Journal of Theoretical Biology*, 108:333–337, 1984.

[27] M. S. Waterman. *Introduction to Computational Biology: Sequences, Maps and Genomes*. Chapman & Hall/CRC, London, 1995.

[28] Z. Zhang, P. Berman, T. Wiehe, and W. Miller. Post-processing long pairwise alignments. *Bioinformatics*, 15(12):1012–1019, 1999.

Joong Chae Na, Kangho Roh, and Kunsoo Park
School of Computer Science and Engineering
Seoul National University
Email: {jcna,khroh,kparkg}@theory.snu.ac.kr

Alberto Apostolico
Dipartimento di Ingegneria dell' Informazione
Università di Padova
and
Department of Computer Sciences
Purdue University
Email: axa@cs.purdue.edu

A Unifying Framework for Seed Sensitivity and its Application to Subset Seeds

GREGORY KUCHEROV, LAURENT NOÉ, AND
MIKHAIL ROYTBERG[1]

ABSTRACT. We propose a general approach to compute the seed sensitivity, that can be applied to different definitions of seeds. It treats separately three components of the seed sensitivity problem – a set of target alignments, an associated probability distribution, and a seed model – that are specified by distinct finite automata. The approach is then applied to a new concept of *subset seeds* for which we propose an efficient automaton construction. Experimental results confirm that sensitive subset seeds can be efficiently designed using our approach, and can then be used in similarity search producing better results than ordinary spaced seeds.

1 Introduction

In the framework of pattern matching and similarity search in biological sequences, seeds specify a class of short sequence motif which, if shared by two sequences, are assumed to witness a potential similarity. Spaced seeds have been introduced several years ago [8, 18] and have been shown to improve significantly the efficiency of the search. One of the key problems associated with spaced seeds is a precise estimation of the sensitivity of the associated search method. This is important for comparing seeds and for choosing most appropriate seeds for a sequence comparison problem to solve.

The problem of seed sensitivity depends on several components. First, it depends on the *seed model* specifying the class of allowed seeds and the way that seeds match (*hit*) potential alignments. In the basic case, seeds are specified by binary words of certain length (*span*), possibly with a constraint on the number of 1's (*weight*). However, different extensions of this basic seed model have been proposed in the literature, such as multi-seed (or multi-hit) strategies [2, 14, 18], seed families [17, 20, 23, 16, 22, 6], seeds over non-binary alphabets [9, 19], vector seeds [4, 6].

The second parameter is the class of *target alignments* that are alignment fragments that one aims to detect. Usually, these are *gapless* alignments of a given

---
[1]part of this work has been done during a visit to LORIA/INRIA in summer 2004.

length. Gapless alignments are easy to model, in the simplest case they are represented by binary sequences in the match/mismatch alphabet. This representation has been adopted by many authors [18, 13, 5, 10, 7, 11]. The binary representation, however, cannot distinguish between different types of matches and mismatches, and is clearly insufficient in the case of protein sequences. In [4, 6], an alignment is represented by a sequence of real numbers that are *scores* of matches or mismatches at corresponding positions. A related, but yet different approach is suggested in [19], where DNA alignments are represented by sequences on the ternary alphabet of match/transition/transversion. Finally, another generalization of simple binary sequences was considered in [15], where alignments are required to be *homogeneous*, i.e. to contain no sub-alignment with a score larger than the entire alignment.

The third necessary ingredient for seed sensitivity estimation is the probability distribution on the set of target alignments. Again, in the simplest case, alignment sequences are assumed to obey a Bernoulli model [18, 10]. In more general settings, Markov or Hidden Markov models are considered [7, 5]. A different way of defining probabilities on binary alignments has been taken in [15]: all homogeneous alignments of a given length are considered equiprobable.

Several algorithms for computing the seed sensitivity for different frameworks have been proposed in the above-mentioned papers. All of them, however, use a common dynamic programming (DP) approach, first brought up in [13].

In the present paper, we propose a general approach to computing the seed sensitivity. This approach subsumes the cases considered in the above-mentioned papers, and allows to deal with new combinations of the three seed sensitivity parameters. The underlying idea of our approach is to specify each of the three components – the seed, the set of target alignments, and the probability distribution – by a separate finite automaton.

A deterministic finite automaton (DFA) that recognizes all alignments matched by given seeds was already used in [7] for the case of ordinary spaced seeds. In this paper, we assume that the set of target alignments is also specified by a DFA and, more importantly, that the probabilistic model is specified by a *probability transducer* – a probability-generating finite automaton equivalent to HMM with respect to the class of generated probability distributions.

We show that once these three automata are set, the seed sensitivity can be computed by a unique general algorithm. This algorithm reduces the problem to a computation of the total weight over all paths in an acyclic graph corresponding to the automaton resulting from the product of the three automata. This computation can be done by a well-known dynamic programming algorithm [21, 12] with the time complexity proportional to the number of transitions of the resulting automaton. Interestingly, all above-mentioned seed sensitivity algorithms considered by different authors can be reformulated as instances of this general algorithm.

In the second part of this work, we study a new concept of *subset seeds* – an extension of spaced seeds that allows to deal with a non-binary alignment alphabet and, on the other hand, still allows an efficient hashing method to locate seeds. For this definition of seeds, we define a DFA with a number of states independent of the size of the alignment alphabet. Reduced to the case of ordinary spaced seeds, this DFA construction gives the same worst-case number of states as the Aho-Corasick DFA used in [7]. Moreover, our DFA has always no more states than the DFA of [7], and has substantially less states on average.

Together with the general approach proposed in the first part, our DFA gives an efficient algorithm for computing the sensitivity of subset seeds, for different classes of target alignments and different probability transducers. In the experimental part of this work, we confirm this by running an implementation of our algorithm in order to design efficient subset seeds for different probabilistic models, trained on real genomic data. We also show experimentally that designed subset seeds allow to find more significant alignments than ordinary spaced seeds of equivalent selectivity.

2 General Framework

Estimating the seed sensitivity amounts to compute the probability for a random word (target alignment), drawn according to a given probabilistic model, to belong to a given language, namely the language of all alignments matched by a given seed (or a set of seeds).

2.1 Target Alignments

Target alignments are represented by words over an alignment alphabet $\mathcal{A}$. In the simplest case, considered most often, the alphabet is binary and expresses a match or a mismatch occurring at each alignment column. However, it could be useful to consider larger alphabets, such as the ternary alphabet of match/transition/transversion for the case of DNA (see [19]). The importance of this extension is even more evident for the protein case ([6]), where different types of amino acid pairs are generally distinguished.

Usually, the set of target alignments is a finite set. In the case considered most often [18, 13, 5, 10, 7, 11], target alignments are all words of a given length n. This set is trivially a regular language that can be specified by a deterministic automaton with $(n+1)$ states. However, more complex definitions of target alignments have been considered (see e.g. [15]) that aim to capture more adequately properties of biologically relevant alignments. In general, we assume that the set of target alignments is a finite regular language $L_T \in \mathcal{A}^*$ and thus can be represented by an acyclic DFA $T = <Q_T, q_T^0, q_T^F, \mathcal{A}, \psi_T>$.

2.2 Probability Assignment

Once an alignment language L_T has been set, we have to define a probability distribution on the words of L_T. We do this using probability transducers.

A probability transducer is a finite automaton without final states in which each transition outputs a *probability*.

DEFINITION 5.1. A *probability transducer* G over an alphabet $\mathcal{A}$ is a 4-tuple $<Q_G, q_G^0, \mathcal{A}, \rho_G>$, where Q_G is a finite set of states, $q_G^0 \in Q_G$ is an initial state, and $\rho_G : Q_G \times \mathcal{A} \times Q_G \to [0,1]$ is a real-valued probability function such that $\forall q \in Q_G, \sum_{q' \in Q_G, a \in \mathcal{A}} \rho_G(q, a, q') = 1$.

A *transition* of G is a triplet $e = <q, a, q'>$ such that $\rho(q, a, q') > 0$. Letter a is called the *label* of e and denoted $label(e)$. A probability transducer G is *deterministic* if for each $q \in Q_G$ and each $a \in \mathcal{A}$, there is at most one transition $<q, a, q'>$. For each path $P = (e_1, ..., e_n)$ in G, we define its *label* to be the word $label(P) = label(e_1)...label(e_n)$, and the associated probability to be the product $\rho(P) = \prod_{i=1}^{n} \rho_G(e_i)$. A path is *initial*, if its start state is the initial state q_G^0 of the transducer G.

DEFINITION 5.2. The *probability* of a word $w \in \mathcal{A}^*$ according to a probability transducer $G = <Q_G, q_G^0, \mathcal{A}, \rho_G>$, denoted $\mathcal{P}_G(w)$, is the sum of probabilities of all initial paths in G with the label w. $\mathcal{P}_G(w) = 0$ if no such path exists. The probability $\mathcal{P}_G(L)$ of a finite language $L \subseteq \mathcal{A}^*$ according a probability transducer G is defined by $\mathcal{P}_G(L) = \sum_{w \in L} \mathcal{P}_G(w)$.

Note that for any n and for $L = \mathcal{A}^n$ (all words of length n), $\mathcal{P}_G(L) = 1$.

Probability transducers can express common probability distributions on words (alignments). Bernoulli sequences with independent probabilities of each symbol [18, 10, 11] can be specified with deterministic one-state probability transducers. In Markov sequences of order k [7, 20], the probability of each symbol depends on k previous symbols. They can therefore be specified by a deterministic probability transducer with at most $|\mathcal{A}|^k$ states.

A Hidden Markov model (HMM) [5] corresponds, in general, to a non-deterministic probability transducer. The states of this transducer correspond to the (hidden) states of the HMM, plus possibly an additional initial state. Inversely, for each probability transducer, one can construct an HMM generating the same probability distribution on words. Therefore, non-deterministic probability transducers and HMMs are equivalent with respect to the class of generated probability distributions. The proofs are straightforward and are omitted due to space limitations.

2.3 Seed Automata and Seed Sensitivity

Since the advent of spaced seeds [8, 18], different extensions of this idea have been proposed in the literature (see Introduction). For all of them, the set of possible alignment fragments matched by a seed (or by a set of seeds) is a finite set, and therefore the set of matched alignments is a regular language. For the original spaced seed model, this observation was used by Buhler et al. [7] who proposed an algorithm for computing the seed sensitivity based on a DFA defining the language of alignments matched by the seed. In this paper, we extend this approach to a general one that allows a uniform computation of seed sensitivity for a wide class of settings including different probability distributions on target alignments, as well as different seed definitions.

Consider a seed (or a set of seeds) π under a given seed model. We assume that the set of alignments L_π matched by π is a regular language recognized by a DFA $S_\pi = <Q_S, q_S^0, Q_S^F, \mathcal{A}, \psi_S>$. Consider a finite set L_T of target alignments and a probability transducer G. Under this assumptions, the sensitivity of π is defined as the conditional probability

$$\frac{\mathcal{P}_G(L_T \cap L_\pi)}{\mathcal{P}_G(L_T)}. \quad (1)$$

An automaton recognizing $L = L_T \cap L_\pi$ can be obtained as the product of automata T and S_π recognizing L_T and L_π respectively. Let $K = <Q_K, q_K^0, Q_K^F, \mathcal{A}, \psi_K>$ be this automaton. We now consider the product W of K and G, denoted $K \times G$, defined as follows.

DEFINITION 5.3. *Given a DFA* $K = <Q_K, q_K^0, Q_K^F, \mathcal{A}, \psi_K>$ *and a probability transducer* $G = <Q_G, q_G^0, \mathcal{A}, \rho_G>$, *the product of* K *and* G *is the probability-weighted automaton* $W = <Q_W, q_W^0, Q_W^F, \mathcal{A}, \rho_W>$ *(for short, PW-automaton) such that*

- $Q_W = Q_K \times Q_G$,
- $q_W^0 = (q_K^0, q_G^0)$,
- $q_W^F = \{(q_K, q_G) | q_K \in Q_K^F\}$,
- $\rho_W((q_K, q_G), a, (q_K', q_G')) = \begin{cases} \rho_G(q_G, a, q_G') & \text{if } \psi_K(q_K, a) = q_K', \\ 0 & \text{otherwise.} \end{cases}$

W can be viewed as a non-deterministic probability transducer with final states. $\rho_W((q_K, q_G), a, (q_K', q_G'))$ is the *probability* of the transition $<(q_K, q_G), a, (q_K', q_G')>$. A path in W is called *full* if it goes from the initial to a final state.

LEMMA 5.4. *Let G be a probability transducer. Let L be a finite language and K be a deterministic automaton recognizing L. Let $W = G \times K$. The probability $\mathcal{P}_G(L)$ is equal to sum of probabilities of all full paths in W.*

Proof. Since K is a deterministic automaton, each word $w \in L$ corresponds to a single accepting path in K and the paths in G labeled w (see Definition 5.1) are in one-to-one correspondence with the full path in W accepting w. By definition, $\mathcal{P}_G(w)$ is equal to the sum of probabilities of all paths in G labeled w. Each such path corresponds to a unique path in W, with the same probability. Therefore, the probability of w is the sum of probabilities of corresponding paths in W. Each such path is a full path, and paths for distinct words w are disjoint. The lemma follows. ∎

2.4 Computing Seed Sensitivity

Lemma 5.4 reduces the computation of seed sensitivity to a computation of the sum of probabilities of paths in a PW-automaton.

LEMMA 5.5. *Consider an alignment alphabet $\mathcal{A}$, a finite set $L_T \subseteq \mathcal{A}^*$ of target alignments, and a set $L_\pi \subseteq \mathcal{A}^*$ of all alignments matched by a given seed π. Let $K = <Q_K, q_t^0, Q_K^F, \mathcal{A}, \psi_Q>$ be an acyclic DFA recognizing the language $L = L_T \cap L_\pi$. Let further $G = <Q_G, q_G^0, \mathcal{A}, \rho>$ be a probability transducer defining a probability distribution on the set L_T. Then $\mathcal{P}_G(L)$ can be computed in time*

$$\mathcal{O}(|Q_G|^2 \cdot |Q_K| \cdot |\mathcal{A}|) \qquad (2)$$

and space

$$\mathcal{O}(|Q_G| \cdot |Q_K|). \qquad (3)$$

Proof. By Lemma 5.4, the probability of L with respect to G can be computed as the sum of probabilities of all full paths in W. Since K is an acyclic automaton, so is W. Therefore, the sum of probabilities of all full paths in W leading to final states q_W^F can be computed by a classical DP algorithm [21] applied to acyclic directed graphs ([12] presents a survey of application of this technique to different bioinformatic problems). The time complexity of the algorithm is proportional to the number of transitions in W. W has $|Q_G| \cdot |Q_K|$ states, and for each letter of $\mathcal{A}$, each state has at most $|Q_G|$ outgoing transitions. The bounds follow. ∎

Lemma 5.5 provides a general approach to compute the seed sensitivity. To apply the approach, one has to define three automata:

- a deterministic acyclic DFA T specifying a set of target alignments over an alphabet $\mathcal{A}$ (e.g. all words of a given length, possibly verifying some additional properties),

- a (generally non-deterministic) probability transducer G specifying a probability distribution on target alignments (e.g. Bernoulli model, Markov sequence of order k, HMM),

- a deterministic DFA S_π specifying the seed model via a set of matched alignments.

As soon as these three automata are defined, Lemma 5.5 can be used to compute probabilities $\mathcal{P}_G(L_T \cap L_\pi)$ and $\mathcal{P}_G(L_T)$ in order to estimate the seed sensitivity according to (1).

Note that if the probability transducer G is deterministic (as it is the case for Bernoulli models or Markov sequences), then the time complexity (2) is $\mathcal{O}(|Q_G| \cdot |Q_K| \cdot |\mathcal{A}|)$. In general, the complexity of the algorithm can be improved by reducing the involved automata. Buhler et al. [7] introduced the idea of using the Aho-Corasick automaton [1] as the seed automaton S_π for a spaced seed. The authors of [7] considered all binary alignments of a fixed length n distributed according to a Markov model of order k. In this setting, the obtained complexity was $\mathcal{O}(w2^{s-w}2^k n)$, where s and w are seed's span and weight respectively. Given that the size of the Aho-Corasick automaton is $\mathcal{O}(w2^{s-w})$, this complexity is automatically implied by Lemma 5.5, as the size of the probability transducer is $\mathcal{O}(2^k)$, and that of the target alignment automaton is $\mathcal{O}(n)$. Compared to [7], our approach explicitly distinguishes the descriptions of matched alignments and their probabilities, which allows us to automatically extend the algorithm to more general cases.

Note that the idea of using the Aho-Corasick automaton can be applied to more general seed models than individual spaced seeds (e.g. to multiple spaced seeds, as pointed out in [7]). In fact, all currently proposed seed models can be described by a finite set of matched alignment fragments, for which the Aho-Corasick automaton can be constructed. We will use this remark in later sections.

The sensitivity of a spaced seed with respect to an HMM-specified probability distribution over binary target alignments of a given length n was studied by Brejova et al. [5]. The DP algorithm of [5] has a lot in common with the algorithm implied by Lemma 5.5. In particular, the states of the algorithm of [5] are triples $< w, q, m >$, where w is a prefix of the seed π, q is a state of the HMM, and $m \in [0..n]$. The states therefore correspond to the construction implied by Lemma 5.5. However, the authors of [5] do not consider any automata, which does not allow to optimize the preprocessing step (counterpart of the automaton construction) and, on the other hand, does not allow to extend the algorithm to more general seed models and/or different sets of target alignments.

A key to an efficient solution of the sensitivity problem remains the definition of the seed. It should be expressive enough to be able to take into account properties of biological sequences. On the other hand, it should be simple enough to be able to locate seeds fast and to get an efficient algorithm for computing seed sensitivity. According to the approach presented in this section, the latter is directly related to the size of a DFA specifying the seed.

3 Subset Seeds

3.1 Definition

Ordinary spaced seeds use the simplest possible binary "match-mismatch" alignment model that allows an efficient implementation by hashing all occurring combinations of matching positions. A powerful generalization of spaced seeds, called *vector seeds*, has been introduced in [4]. Vector seeds allow one to use an arbitrary alignment alphabet and, on the other hand, provide a flexible definition of a hit based on a cooperative contribution of seed positions. A much higher expressiveness of vector seeds lead to more complicated algorithms and, in particular, prevents the application of direct hashing methods at the seed location stage.

In this section, we consider *subset seeds* that have an intermediate expressiveness between spaced and vector seeds. It allows an arbitrary alignment alphabet and, on the other hand, still allows using a direct hashing for locating seed, which maps each string to a unique entry of the hash table. We also propose a construction of a seed automaton for subset seeds, different from the Aho-Corasick automaton. The automaton has $\mathcal{O}(w2^{s-w})$ states *regardless of the size of the alignment alphabet*, where s and w are respectively the span of the seed and the number of "must-match" positions. From the general algorithmic framework presented in the previous section (Lemma 5.5), this implies that the seed sensitivity can be computed for subset seeds with same complexity as for ordinary spaced seeds. Note also that for the binary alignment alphabet, this bound is the same as the one implied by the Aho-Corasick automaton. However, for larger alphabets, the Aho-Corasick construction leads to $\mathcal{O}(w|\mathcal{A}|^{s-w})$ states. In the experimental part of this paper (section 4.1) we will show that even for the binary alphabet, our automaton construction yields a smaller number of states in practice.

Consider an alignment alphabet $\mathcal{A}$. We always assume that $\mathcal{A}$ contains a symbol 1, interpreted as "match". A *subset seed* is defined as a word over a *seed alphabet* $\mathcal{B}$, such that

- letters of $\mathcal{B}$ denote subsets of the alignment alphabet $\mathcal{A}$ containing 1 ($\mathcal{B} \subseteq \{1\} \cup 2^{\mathcal{A}}$),

- $\mathcal{B}$ contains a letter $\#$ that denotes subset $\{1\}$,

- a subset seed $b_1 b_2 \ldots b_m \in \mathcal{B}^m$ matches an alignment fragment $a_1 a_2 \ldots a_m \in \mathcal{A}^m$ if $\forall i \in [1..m]$, $a_i \in b_i$.

The $\#$-*weight* of a subset seed π is the number of $\#$ in π and the *span* of π is its length.

EXAMPLE 5.6. [19] considered the alignment alphabet $\mathcal{A} = \{1, \mathtt{h}, \mathtt{0}\}$ representing respectively a match, a transition mismatch, or a transversion mismatch in a

DNA sequence alignment. The seed alphabet is $\mathcal{B} = \{\#, @, \_\}$ denoting respectively subsets $\{1\}, \{1, h\}$, and $\{1, h, 0\}$. Thus, seed $\pi = \#@\_\#$ matches alignment $s = $ 10h1h1101 at positions 4 and 6. The span of π is 4, and the #-weight of π is 2.

Note that unlike the weight of ordinary spaced seeds, the #-weight cannot serve as a measure of seed selectivity. In the above example, symbol @ should be assigned weight 0.5, so that the weight of π is equal to 2.5 (see [19]).

3.2 Subset Seed Automaton

Let us fix an alignment alphabet $\mathcal{A}$, a seed alphabet $\mathcal{B}$, and a seed $\pi = \pi_1 \pi_2 \ldots \pi_m \in \mathcal{B}^*$ of span m and #-weight w. Let R_π be the set of all non-# positions in π, $|R_\pi| = r = m - w$. We now define an automaton $S_\pi = <Q, q_0, Q_f, \mathcal{A}, \psi : Q \times \mathcal{A} \to Q>$ that recognizes the set of all alignments matched by π.

The states Q of S_π are pairs $<X, t>$ such that $X \subseteq R_\pi, t \in [0, \ldots, m]$, with the following invariant condition. Suppose that S_π has read a prefix $s_1 \ldots s_p$ of an alignment s and has come to a state $<X, t>$. Then t is the length of the longest suffix of $s_1 \ldots s_p$ of the form 1^i, $i \leq m$, and X contains all positions $x_i \in R_\pi$ such that prefix $\pi_1 \cdots \pi_{x_i}$ of π matches a suffix of $s_1 \cdots s_{p-t}$.

(a)

$\pi = \#@\#\_\#\#\_\#\#\#$

(b)

$s = $ 111h1011h11...

(c)

$\overset{s_9\ \ t}{\text{111h1011h11}\ldots}$

$\pi_{1..7} = \#@\#\_\#\#\_$
$\pi_{1..4} = \#@\#\_$
$\pi_{1..2} = \#@$

Figure 1. Illustration to Example 5.7.

EXAMPLE 5.7. In the framework of Example 5.6, consider a seed π and an alignment prefix s of length $p = 11$ given on Figure 1(a) and (b) respectively. The length t of the last run of 1's of s is 2. The last mismatch position of s is $s_9 = $ h. The set R_π of non-# positions of π is $\{2, 4, 7\}$ and π has 3 prefixes ending at positions of R_π (Figure 1(c)). Prefixes $\pi_{1..2}$ and $\pi_{1..7}$ do match suffixes of $s_1 s_2 \ldots s_9$, and prefix $\pi_{1..4}$ does not. Thus, the state of the automaton after reading $s_1 s_2 \ldots s_{11}$ is $<\{2, 7\}, 2>$.

The initial state q_0 of S_π is the state $<\emptyset, 0>$. The final states Q_f of S_π are all states $q = <X, t>$, where $max\{X\} + t = m$. All final states are merged into one state.

The transition function $\psi(q, a)$ is defined as follows: If q is a final state, then $\forall a \in \mathcal{A}, \psi(q, a) = q$. If $q = <X, t>$ is a non-final state, then

- if $a = 1$ then $\psi(q,a) = <X, t+1>$,
- otherwise $\psi(q,a) = <X_U \cup X_V, 0>$ with
  - $X_U = \{x | x \leq t+1 \text{ and } a \text{ matches } \pi_x\}$
  - $X_V = \{x+t+1 | x \in X \text{ and } a \text{ matches } \pi_{x+t+1}\}$

LEMMA 5.8. *The automaton S_π accepts the set of all alignments matched by π.*

Proof. It can be verified by induction that the invariant condition on the states $<X, t> \in Q$ is preserved by the transition function ψ. The final states verify $max\{X\} + t = m$, which implies that π matches a suffix of $s_1 \ldots s_p$. ∎

LEMMA 5.9. *The number of states of the automaton S_π is no more than $(w+1)2^r$.*

Proof. Assume that $R_\pi = \{x_1, x_2, \ldots, x_r\}$ and $x_1 < x_2 \cdots < x_r$. Let Q_i be the set of non-final states $<X, t>$ with $max\{X\} = x_i$, $i \in [1..r]$. For states $q = <X, t> \in Q_i$ there are 2^{i-1} possible values of X and $m - x_i$ possible values of t, as $max\{X\} + t \leq m - 1$.
Thus,

$$|Q_i| \leq 2^{i-1}(m - x_i) \leq 2^{i-1}(m - i), \text{ and} \quad (4)$$

$$\sum_{i=1}^{r} |Q_i| \leq \sum_{i=1}^{r} 2^{i-1}(m-i) = (m - r + 1)2^r - m - 1. \quad (5)$$

Besides states Q_i, Q contains m states $<\emptyset, t>$ ($t \in [0..m-1]$) and one final state. Thus, $|Q| \leq (m - r + 1)2^r = (w + 1)2^r$. ∎

Note that if π starts with #, which is always the case for ordinary spaced seeds, then $X_i \geq i + 1$, $i \in [1..r]$, and the bound of (4) rewrites to $2^{i-1}(m - i - 1)$. This results in the same number of states $w2^r$ as for the Aho-Corasick automaton [7]. The construction of automaton S_π is optimal, in the sense that no two states can be merged in general, as the following Lemma states.

LEMMA 5.10. *Consider a spaced seed π which consists of two "must-match" symbols # separated by r jokers. Then the automaton S_π is reduced, that is any non-final state is reachable from the initial state q_0, and any two non-final states q, q' are non-equivalent.*

Proof. See appendix A. ∎

A straightforward generation of the transition table of the automaton S_π can be performed in time $\mathcal{O}(r \cdot w \cdot 2^r \cdot |\mathcal{A}|)$. A more complicated algorithm allows one to reduce the bound to $\mathcal{O}(w \cdot 2^r \cdot |\mathcal{A}|)$. This algorithm is described in full details in Appendix B. Here we summarize it in the following Lemma.

LEMMA 5.11. *The transition table of automaton S_π can be constructed in time proportional to its size, which is $\mathcal{O}(w \cdot 2^r \cdot |\mathcal{A}|)$.*

In the next section, we demonstrate experimentally that on average, our construction yields a very compact automaton, close to the minimal one. Together with the general approach of section 2, this provides a fast algorithm for computing the sensitivity of subset seeds and, in turn, allows to perform an efficient design of spaced seeds well-adapted to the similarity search problem under interest.

4 Experiments

Several types of experiments have been performed to test the practical applicability of the results of sections 2,3. We focused on DNA similarity search, and set the alignment alphabet $\mathcal{A}$ to $\{1, h, 0\}$ (match, transition, transversion). For subset seeds, the seed alphabet $\mathcal{B}$ was set to $\{\#, @, \_\}$, where $\# = \{1\}, @ = \{1, h\}, \_ = \{1, h, 0\}$ (see Example 5.6). The weight of a subset seed is computed by assigning weights 1, 0.5 and 0 to symbols $\#$, $@$ and $\_$ respectively.

4.1 Size of the Automaton

We compared the size of the automaton S_π defined in section 3 and the Aho-Corasick automaton [1], both for ordinary spaced seeds (binary seed alphabet) and for subset seeds. The Aho-Corasick automaton for spaced seeds was constructed as defined in [7]. For subset seeds, a straightforward generalization was considered: the Aho-Corasick construction was applied to the set of alignment fragments matched by the seed.

Tables 1(a) and 1(b) present the results for spaced seeds and subset seeds respectively. For each seed weight w, we computed the average number of states ($avg.\ size$) of the Aho-Corasick automaton and our automaton S_π, and reported the corresponding ratio (δ) with respect to the average number of states of the minimized automaton. The average was computed over all seeds of span up to $w + 8$ for spaced seeds and all seeds of span up to $w + 5$ with two @'s for subset seeds. Interestingly, our automaton turns out to be more compact than the Aho-Corasick automaton not only on non-binary alphabets (which was expected), but also on the binary alphabet (cf Table 1(a)). Note that for a given seed, one can define a surjective mapping from the states of the Aho-Corasick automaton onto the states of our automaton. This implies that our automaton has *always* no more states than the Aho-Corasick automaton.

4.2 Seed Design

In this part, we considered several probability transducers to design spaced or subset seeds. The target alignments included all alignments of length 64 on alphabet

Table 1. *Comparison of the average number of states of Aho-Corasick automaton, automaton S_π of section 3 and minimized automaton.*

Spaced	Aho-Corasick		S_π		Minimized
w	avg. size	δ	avg. size	δ	avg. size
9	345.94	3.06	146.28	1.29	113.21
10	380.90	3.16	155.11	1.29	120.61
11	415.37	3.25	163.81	1.28	127.62
12	449.47	3.33	172.38	1.28	134.91
13	483,27	3.41	180.89	1.28	141.84

(a)

Subset	Aho-Corasick		S_π		Minimized
w	avg. size	δ	avg. size	δ	avg. size
9	1900.65	15.97	167.63	1.41	119,00
10	2103.99	16.50	177.92	1.40	127.49
11	2306.32	16.96	188.05	1.38	135.95
12	2507.85	17.42	198.12	1.38	144.00
13	2709.01	17.78	208.10	1.37	152.29

(b)

$\{1, h, 0\}$. Four probability transducers have been studied (analogous to those introduced in [3]):

- B: Bernoulli model

- $DT1$: deterministic probability transducer specifying probabilities of $\{1, h, 0\}$ at each codon position (extension of the $M^{(3)}$ model of [3] to the three-letter alphabet),

- $DT2$: deterministic probability transducer specifying probabilities of each of the 27 codon instances $\{1, h, 0\}^3$ (extension of the $M^{(8)}$ model of [3] to the three-letter alphabet),

- NT: non-deterministic probability transducer combining four copies of $DT2$

specifying four distinct codon conservation levels (called HMM model in [3]).

Models $DT1$, $DT2$ and NT have been trained on alignments resulting from a pairwise comparison of 40 bacteria genomes. Details of the training procedure as well as the resulting parameter values are given in Appendix C.

For each of the four probability transducers, we computed the best seed of weight w ($w = 9, 10, 11, 12$) among two categories: ordinary spaced seeds of weight w and subset seeds of weight w with two @. Ordinary spaced seeds were enumerated exhaustively up to a given span, and for each seed, the sensitivity was computed using the algorithmic approach of section 2 and the seed automaton construction of section 3. Each such computation took between 10 and 500ms on a Pentium IV 2.4GHz computer depending on the seed weight/span and the model used. In each experiment, the most sensitive seed found has been kept. The results are presented in Tables 2-5.

Table 2. *Best seeds and their sensitivity for probability transducer B.*

w	spaced seeds	Sens.	subset seeds, two @	Sens.
9	###_ _ _#_#_##_##	0.4183	###_#_ _#@#_@##	0.4443
10	##_##_ _ _##_#_###	0.2876	###_@#_@#_#_###	0.3077
11	###_###_#_ _#_###	0.1906	##@#_ _##_#_#_@###	0.2056
12	###_#_##_#_ _##_###	0.1375	##@#_#_##_ _#@_####	0.1481

Table 3. *Best seeds and their sensitivity for probability transducer DT1.*

w	spaced seeds	Sens.	subset seeds, two @	Sens.
9	###_ _ _##_##_##	0.4350	##@_ _ _##_##_##@	0.4456
10	##_##_ _ _ _##_##_##	0.3106	##_##_ _ _@##_##@#	0.3173
11	##_##_ _ _ _##_##_###	0.2126	##@#@_##_##_ _###	0.2173
12	##_##_ _ _ _##_##_####	0.1418	##_@###_ _##_##@##	0.1477

In all cases, subset seeds yield a better sensitivity than ordinary spaced seeds. The sensitivity increment varies up to 0.04 which is a notable increase. As shown in [19], the gain in using subset seeds increases substantially when the transition

Table 4. *Best seeds and their sensitivity for probability transducer DT2.*

w	spaced seeds	Sens.	subset seeds, two @	Sens.
9	#_##____##_##_##	0.5121	#_#@_##_@__##_##	0.5323
10	##_##_##____##_##	0.3847	##_@#_##__@_##_##	0.4011
11	##_##__#_#___#_##_##	0.2813	##_##_@#_#___#_#@_##	0.2931
12	##_##_##_#___#_##_##	0.1972	##_##_#@_##_@__##_##	0.2047

Table 5. *Best seeds and their sensitivity for probability transducer NT.*

w	spaced seeds	Sens.	subset seeds, two @	Sens.
9	##_##_##____##_#	0.5253	##_@@_##____##_##	0.5420
10	##_##____##_##_##	0.4123	##_##____##_@@_##_#	0.4190
11	##_##____##_##_##_#	0.3112	##_##____##_@@_##_##	0.3219
12	##_##____##_##_##_##	0.2349	##_##____##_@@_##_##_#	0.2412

probability is greater than the inversion probability, which is very often the case in related genomes.

4.3 Comparative Performance of Spaced and Subset Seeds

We performed a series of whole genome comparisons in order to compare the performance of designed spaced and subset seeds. Eight complete bacterial genomes[2] have been processed against each other using the YASS software [19]. Each comparison was done twice: one with a spaced seed and another with a subset seed of the same weight.

The threshold E-value for the output alignments was set to 10, and for each comparison, the number of alignments with E-value smaller than 10^{-3} found by each seed, and the number of exclusive alignments were reported. By "exclusive alignment" we mean any alignment of E-value less than 10^{-3} that does not share a common part (do not overlap in both compared sequences) with any alignment found by another seed. To take into account a possible bias caused by splitting alignments into smaller ones (X-drop effect), we also computed the total length of exclusive alignments. Table 6 summarizes these experiments for weights 9 and 10

---

[2]NC_000907.fna, NC_002662.fna, NC_003317.fna, NC_003454.fna, NC_004113.fna, NC_001263.fna, NC_003112.fna obtained from NCBI

and the $DT2$ and NT probabilistic models. Each line corresponds to a seed given in Table 4 or Table 5, depending on the indicated probabilistic model.

Table 6. *Comparative test of subset seeds vs spaced seeds. Reported execution times (min:sec) were obtained on a Pentium IV 2.4GHz computer.*

seed	time	#align	#ex.align	ex. align length
$DT2, w = 9$, spaced seed	15:14	19101	1583	130512
$DT2, w = 9$, subset seed, two @	14:01	20127	1686	141560
$DT2, w = 10$, spaced seed	8:45	18284	1105	10174
$DT2, w = 10$, subset seed, two @	8:27	18521	1351	12213
$NT, w = 9$, spaced seed	42:23	20490	1212	136049
$NT, w = 9$, subset seed, two @	41:58	21305	1497	150127
$NT, w = 10$, spaced seed	11:45	19750	942	85208
$NT, w = 10$, subset seed, two @	10:31	21652	1167	91240

In all cases, best subset seeds detect from 1% to 8% more significant alignments compared to best spaced seeds of same weight.

5 Discussion

We introduced a general framework for computing the seed sensitivity for various similarity search settings. The approach can be seen as a generalization of methods of [7, 5] in that it allows to obtain algorithms with the same worst-case complexity bounds as those proposed in these papers, but also allows to obtain efficient algorithms for new formulations of the seed sensitivity problem. This versatility is achieved by distinguishing and treating separately the three ingredients of the seed sensitivity problem: a set of target alignments, an associated probability distributions, and a seed model.

We then studied a new concept of *subset seeds* which represents an interesting compromise between the efficiency of spaced seeds and the flexibility of vector seeds. For this type of seeds, we defined an automaton with $\mathcal{O}(w2^r)$ states regardless of the size of the alignment alphabet, and showed that its transition table can be constructed in time $\mathcal{O}(w2^r|\mathcal{A}|)$. Projected to the case of spaced seeds, this construction gives the same worst-case bound as the Aho-Corasick automaton of [7], but results in a smaller number of states in practice. Different experiments we have done confirm the practical efficiency of the whole method, both at the level of computing sensitivity for designing good seeds, as well as using those seeds for DNA similarity search.

As far as the future work is concerned, it would be interesting to study the design of efficient spaced seeds for protein sequence search (see [6]), as well as to combine spaced seeds with other techniques such as seed families [17, 20, 16] or the group hit criterion [19].

Acknowledgements

G. Kucherov and L. Noé have been supported by the *ACI IMPBio* of the French Ministry of Research. A part of this work has been done during a stay of M. Roytberg at LORIA, Nancy, supported by INRIA. M.Roytberg has been also supported by the Russian Foundation for Basic Research (projects 03-04-49469, 02-07-90412) and by grants from the RF Ministry of Industry, Science and Technology (20/2002, 5/2003) and NWO (Netherlands Science Foundation).

BIBLIOGRAPHY

[1] AHO, A. V., AND CORASICK, M. J. Efficient string matching: An aid to bibliographic search. *Communications of the ACM 18*, 6 (1975), 333–340.

[2] ALTSCHUL, S., MADDEN, T., SCHÄFFER, A., ZHANG, J., ZHANG, Z., MILLER, W., AND LIPMAN, D. Gapped BLAST and PSI-BLAST: a new generation of protein database search programs. *Nucleic Acids Research 25*, 17 (1997), 3389–3402.

[3] BREJOVA, B., BROWN, D., AND VINAR, T. Optimal spaced seeds for hidden markov models, with application to homologous coding regions. In *Proceedings of the 14th Symposium on Combinatorial Pattern Matching, Morelia (Mexico)* (June 2003), M. C. R. Baeza-Yates, E. Chavez, Ed., vol. 2676 of *Lecture Notes in Computer Science*, Springer, pp. 42–54.

[4] BREJOVA, B., BROWN, D., AND VINAR, T. Vector seeds: an extension to spaced seeds allows substantial improvements in sensitivity and specificity. In *Proceedings of the 3rd International Workshop in Algorithms in Bioinformatics (WABI), Budapest (Hungary)* (September 2003), G. Benson and R. Page, Eds., vol. 2812 of *Lecture Notes in Computer Science*, Springer.

[5] BREJOVA, B., BROWN, D., AND VINAR, T. Optimal spaced seeds for homologous coding regions. *Journal of Bioinformatics and Computational Biology 1*, 4 (Jan 2004), 595–610.

[6] D. BROWN. Optimizing multiple seeds for protein homology search. *IEEE Transactions on Computational Biology and Bioinformatics 2*, 1 (Jan. 2005), 29 – 38.

[7] BUHLER, J., KEICH, U., AND SUN, Y. Designing seeds for similarity search in genomic dna. In *Proceedings of the 7th Annual International Conference on Computational Molecular Biology (RECOMB03), Berlin (Germany)* (April 2003), ACM Press, pp. 67–75.

[8] BURKHARDT, S., AND KÄRKKÄINEN, J. Better filtering with gapped q-grams. *Fundamenta Informaticae 56*, 1-2 (2003), 51–70. Preliminary version in Combinatorial Pattern Matching 2001.

[9] CHEN, W., AND SUNG, W.-K. On half gapped seed. *Genome Informatics 14* (2003), 176–185. preliminary version in the 14th International Conference on Genome Informatics (GIW).

[10] CHOI, K., AND ZHANG, L. Sensitivity analysis and efficient method for identifying optimal spaced seeds. *Journal of Computer and System Sciences 68* (2004), 22–40.

[11] CHOI, K. P., ZENG, F., AND ZHANG, L. Good spaced seeds for homology search. *Bioinformatics 20* (2004), 1053–1059.

[12] FINKELSTEIN, A., AND ROYTBERG, M. Computation of biopolymers: A general approach to different problems. *BioSystems 30*, 1-3 (1993), 1–19.

[13] KEICH, U., LI, M., MA, B., AND TROMP, J. On spaced seeds for similarity search. to appear in Discrete Applied Mathematics, 2002.

[14] KENT, W. J. Blat–the blast-like alignment tool. *Genome Research 12* (2002), 656–664.

[15] KUCHEROV, G., NOÉ, L., AND PONTY, Y. Estimating seed sensitivity on homogeneous alignments. In *Proceedings of the IEEE 4th Symposium on Bioinformatics and Bioengineering (BIBE 2004), May 19-21, 2004, Taichung (Taiwan)* (2004), IEEE Computer Society Press, pp. 387–394.

[16] KUCHEROV, G., NOÉ, L., AND ROYTBERG, M. Multiseed lossless filtration. *IEEE Transactions on Computational Biology and Bioinformatics 2*, 1 (Jan. 2005), 51 – 61.

[17] LI, M., MA, B., KISMAN, D., AND TROMP, J. Patternhunter ii: Highly sensitive and fast homology search. *Journal of Bioinformatics and Computational Biology* (2004). Earlier version in GIW 2003 (International Conference on Genome Informatics).

[18] MA, B., TROMP, J., AND LI, M. Patternhunter: Faster and more sensitive homology search. *Bioinformatics 18*, 3 (2002), 440–445.

[19] NOÉ, L., AND KUCHEROV, G. Improved hit criteria for dna local alignment. *BMC Bioinformatics 5*, 149 (14 October 2004).

[20] SUN, Y., AND BUHLER, J. Designing multiple simultaneous seeds for DNA similarity search. In *Proceedings of the 8th Annual International Conference on Computational Molecular Biology (RECOMB04), San Diego (California)* (March 2004), ACM Press.

[21] ULLMAN, J. D., AHO, A. V., AND HOPCROFT, J. E. *The Design and Analysis of Computer Algorithms*. Addison-Wesley, Reading, 1974.

[22] XU, J., BROWN, D., LI, M., AND MA, B. Optimizing multiple spaced seeds for homology search. In *Proceedings of the 15th Symposium on Combinatorial Pattern Matching, Istambul (Turkey)* (July 2004), vol. 3109 of *Lecture Notes in Computer Science*, Springer.

[23] YANG, I.-H., WANG, S.-H., CHEN, Y.-H., HUANG, P.-H., YE, L., HUANG, X., AND CHAO, K.-M. Efficient methods for generating optimal single and multiple spaced seeds. In *Proceedings of the IEEE 4th Symposium on Bioinformatics and Bioengineering (BIBE 2004), May 19-21, 2004, Taichung (Taiwan)* (2004), IEEE Computer Society Press, pp. 411–416.

A Proof of Lemma 5.10

Let $\pi = \# -^r \#$ be a spaced seed of span $r+2$ and weight 2. We prove that the automaton S_π (see Lemma 5.8) is reduced, i.e.

(i) all its non-final states are reachable from the initial state $< \emptyset, 0 >$;

(ii) any two non-final states q, q' are non-equivalent, i.e. there is a word $w = w(q, q')$ such that exactly one of the states $\psi(q, w), \psi(q', w)$ is a final state.

(i) Let $q = <X, t>$ be a state of the automaton S_π, and let $X = \{x_1, \ldots, x_k\}$ and $x_1 < \cdots < x_k$. Obviously, $x_k + t < r+2$. Let $s \in \{0,1\}^*$ be an alignment word of length x_k such that for all $i \in [1, x_k]$, $s_i = 1$ iff $\exists j \in [1, k]$, $i = x_k - x_j + 1$. Note, that, $\pi_1 = \#$, therefore $1 \notin X$ and $s_{x_k} = 0$. Finally, $\psi(<\phi, 0>, s \cdot 1^t) = q$.

(ii) Let $q_1 = <X_1, t_1>$ and $q_2 = <X_2, t_2>$ be non-final states of S_π. Let $X_1 = \{y_1, \ldots, y_a\}, X_2 = \{z_1, \ldots, z_b\}$, and $y_1 < \cdots < y_a, z_1 < \cdots < z_b$.

Assume that $max\{X_1\} + t_1 > max\{X_2\} + t_2$ and let $d = (r+2) - (max\{X_1\} + t_1)$. Obviously, $\psi(q_1, 1^d)$ is a final state, and $\psi(q_2, 1^d)$ is not. Now assume that $max\{X_1\} + t_1 = max\{X_2\} + t_2$. For a set $X \subseteq \{1, \ldots, r+1\}$ and a number t, define a set $X\{t\}$ by $X\{t\} = \{v + t | v \in X$ and $v + t < r+2\}$. Let $g = max\{v | (v + t_1 \in X_1$ and $v + t_2 \notin X_2)$ or $(v + t_2 \in X_2$ and $v + t_1 \notin X_1)\}$ and let $d = r+1-g$. Then $\psi(q_1, 0^d \cdot 1)$ is a final state and $\psi(q_2, 0^d \cdot 1)$ is not or vice versa. This completes the proof.

B Subset Seed Automaton

Let π be a subset seed of $\#$-weight w and span s, and $r = s - w$ be the number of non-$\#$ positions. We define a DFA S_π recognizing all words of $\mathcal{A}^*$ matched by π

(see definition of section 3.1). The transition table of S_π is stored in an array such that each element describes a state $<X,t>$ of S_π. Now we define

1. how to compute the array index $Ind(q)$ of a state $q = <X,t>$,

2. how to compute values $\psi(q,a)$ given a state q and a letter $a \in \mathcal{A}$.

B.1 Encoding State Indexes

We will need some notation. Let $L = \{l_1, \ldots, l_r\}$ be a set of all non-# positions in π ($l_1 < l_2 < \cdots < l_r$). For a subset $X \subseteq L$, let $v(X) = v_1 \ldots v_r \in \{0,1\}^r$ be a binary vector such that $v_i = 1$ iff $l_i \in X$. Let further $n(X)$ be the integer corresponding to the binary representation $v(X)$ (read from left to right):

$$n(X) = \sum_{j=1}^{r} 2^{j-1} \cdot v_j.$$

Define $p(t) = max\{p | l_p < m-t\}$. Informally, for a given non-final state $<X,t>$, X can only be a subset of $\{l_1, \ldots, l_{p(t)}\}$. This implies that $n(X) < 2^{p(t)}$. Then, the index of a given state $\{<X,t>\}$ in the array is defined by

$$Ind(<X,t>) = n(X) + 2^{p(t)}.$$

This implies that the worst-case size of the array is no more than $w2^r$ (the proof is similar to the proof of Lemma 5.9).

B.2 Computing Transition Function $\psi(q,a)$

We compute values $\psi(<X,t>,a)$ based on already computed values $\psi(<X',t>,a)$. Let $q = <X,t>$ be a non-final and reachable state of S_π, where $X = \{l_1, \ldots, l_k\}$ with $l_1 < l_2 \cdots < l_k$ and $k \leq r$. Let $X' = X \setminus \{l_k\} = \{l_1, \ldots, l_{k-1}\}$ and $q' = <X',t>$. Then the following lemma holds.

LEMMA .12. *If $q = <X,t>$ is reachable, then $q' = <X',t>$ is reachable and has been processed before.*

Proof. First prove that $<X',t>$ is reachable. If $<X,t>$ is reachable, then $<X,0>$ is reachable due to the definition of transition function for $t > 0$. Thus, one can find at least one sequence $S \in \mathcal{A}^{l_k}$ such that $\forall i \in [1..r]$, $l_i \in X$ iff $\pi_1 \cdots \pi_{l_i}$ matches $S_{l_k - l_i + 1} \cdots S_{l_k}$. For such a sequence S, one can find a word $S' = S_{l_k - l_{k-1} + 1} \cdots S_{l_k}$ which reaches state $<X',0>$. To conclude, if there exists a word $S \cdot 1^t$ that reaches the state $<X,t>$, there also exists a word $S' \cdot 1^t$ that reaches $<X',t>$.

Note that as $|S' \cdot 1^t| < |S \cdot 1^t|$, then a breadth-first computation of states of S_π always processes state $<X',t>$ before $<X,t>$. ∎

Now we present how to compute values $\psi(<X,t>,a)$ from values $\psi(<X',t>, a)$. This is done by Algorithm 1 shown below, that we comment on now. Due to implementation choices, we represent a state q as triple $q = \langle X, k_X, t\rangle$, where $k_X = max\{i|l_i \in X\}$. Note first that if $a = 1$, the transition function $\psi(q,a)$ can be computed in constant time due to its definition (part a. of Algorithm 1). If $a \neq 1$, we have to

1. retrieve the index of q' given $q = \langle X, k_X, t\rangle$ (part c. of Algorithm 1),

2. compute $\psi(\langle X, k_X, t\rangle, a \neq 1)$ given $\psi(\langle X', k_{X'}, t\rangle, a \neq 1)$ value. (part d. of Algorithm 1)

1. Note first that $Ind(\langle X, k_X, t\rangle) = Ind(\langle X', k_{X'}, t\rangle) - 2^{k_X}$, which can be computed in constant time since k_X is explicitly stored in the current state.
2. Let

$$V_X(k, t, a \neq 1) = \begin{cases} l_i & \text{if } l_i = l_k + t + 1 \text{ and } a \text{ matches } \pi_{l_i} \\ \emptyset & \text{otherwise} \end{cases}$$

and

$$V_k(k, t, a \neq 1) = \begin{cases} i & \text{if } l_i = l_k + t + 1 \text{ and } a \text{ matches } \pi_{l_i} \\ 0 & \text{otherwise} \end{cases}$$

Tables $V_X(k, t, a)$ and $V_k(k, t, a)$ can be precomputed in time and space $\mathcal{O}(|\mathcal{A}| \cdot m^2)$. Let $\psi(\langle X, k_X, t\rangle, a) = \langle Y, k_Y, 0\rangle$ and $\psi(\langle X', k_{X'}, t\rangle, a) = \langle Y', k_{Y'}, 0\rangle$. The set Y differs from Y' at most with one element. This element can be computed in constant time using tables V_X, V_k. Namely $Y = Y' \cup V_X(k_X, t, a)$ and $k_Y = max(k_{Y'}, V_k(k_X, t, a))$.

Note that a final situation arises when $X = \emptyset$. (part b. of Algorithm 1). One also has to compute two tables U_X, U_k defined as:

$$U_X(t, a \neq 1) = \cup\{x | x \leq t + 1 \text{ and } a \text{ matches } \pi_x\}$$
$$U_k(t, a \neq 1) = max\{x | x \leq t + 1 \text{ and } a \text{ matches } \pi_x\}$$

LEMMA .13. *The transition function $\psi(q, a)$ can be computed in constant time for every reachable state q and every $a \in \mathcal{A}$.*

Data: a seed π of span m, $'\#'$-weight w, and number of jokers $r = m - w$
Result: an automaton $S_\pi = <Q, q_0, q_F, \mathcal{A}, \psi>$
$Q.add(q_F)$;
$q_0 \leftarrow \langle X = \emptyset, k = 0, t = 0 \rangle$;
$Q.add(q_0)$;
$queue.push(q_0)$;
while $queue \neq \emptyset$ **do**
    $\langle X, k_X, t_X \rangle = queue.pop()$;
    for $a \in \mathcal{A}$ **do**
        /* compute $\psi(<X, t_X>, a) = \langle Y, k_Y, t_Y \rangle$ */
        if $a = 1$ **then**

a

            $t_Y \leftarrow t_X + 1$;
            $k_Y \leftarrow k_X$;
            $Y \leftarrow X$;
        else
            if $X = \emptyset$ **then**

b

                $Y \leftarrow U_X(t_X, a)$;
                $k_Y \leftarrow U_k(t_X, a)$;
            else
                /* use already processed $\psi(<X', t_{X'}>, a)$ ... */

c

                $X' \leftarrow X \setminus \{l_{k_X}\}$;
                $\langle Y', k_{Y'}, t_{Y'} \rangle \leftarrow \psi(<X', t>, a)$;
                /* ... to compute $\psi(<X, t_X>, a)$ */

d

                $k_Y \leftarrow max(k_{Y'}, V_k(k_X, t_X, a))$;
                $Y \leftarrow Y' \cup V_X(k_X, t_X, a)$;
            $t_Y \leftarrow 0$;
        if $L[k_Y] + t_Y \geq m$ **then**
            /* $<Y, t_Y>$ is a final state */
            $\psi(<X, t_X>, a) \leftarrow q_F$;
        else
            if $\langle Y, k_Y, t_Y \rangle \notin Q$ **then**
                $Q.add(\langle Y, k_Y, t_Y \rangle)$;
                $queue.push(\langle Y, k_Y, t_Y \rangle)$;
            $\psi(<X, t_X>, a) \leftarrow \langle Y, k_Y, t_Y \rangle$;

Algorithm 1: S_π computation

C Training Probability Transducers

We selected 40 bacterial complete genomes from NCBI: NC_000117.fna, NC_000907.fna, NC_000909.fna, NC_000922.fna, NC_000962.fna, NC_001263.fna, NC_001318.fna, NC_002162.fna, NC_002488.fna, NC_002505.fna, NC_002516.fna, NC_002662.fna, NC_002678.fna, NC_002696.fna, NC_002737.fna, NC_002927.fna, NC_003037.fna, NC_003062.fna, NC_003112.fna, NC_003210.fna, NC_003295.fna, NC_003317.fna, NC_003454.fna, NC_003551.fna, NC_003869.fna, NC_003995.fna, NC_004113.fna, NC_004307.fna, NC_004342.fna, NC_004551.fna, NC_004631.fna, NC_004668.fna, NC_004757.fna, NC_005027.fna, NC_005061.fna, NC_005085.fna, NC_005125.fna, NC_005213.fna, NC_005303.fna, NC_005363.fna .

YASS [19] has been run on each pair of genomes to detect alignments with E-value at most 10^{-3}. Resulting ungapped regions of length 64 or more have been used to train models $DT1$, $DT2$ and NT by the maximal likelihood criterion. Table 7 gives the ρ function of the probability transducer $DT1$, that specifies the probabilities of match (1), transition (h) and transversion (0) at each codon position.

Table 7. *Parameters of the $DT1$ model.*

a :	0	h	1
$\rho(q_0, a, q_1)$	0.2398	0.2945	0.4657
$\rho(q_1, a, q_2)$	0.1351	0.1526	0.7123
$\rho(q_2, a, q_0)$	0.1362	0.1489	0.7150

Table 8 specifies the probability of each codon instance $a_1 a_2 a_3 \in \mathcal{A}^3$, used to define the probability transducer $DT2$.

Table 8. *Probability of each codon instance specified by the $DT2$ model.*

$a_1 \backslash$ $a_2 a_3$:	00	0h	01	h0	hh	h1	10	1h	11
0	0.01089	0.01329	0.01311	0.01107	0.00924	0.01144	0.01887	0.01946	0.03106
h	0.01022	0.00984	0.01093	0.00956	0.01025	0.01294	0.02155	0.02552	0.03983
1	0.02083	0.02158	0.02554	0.02537	0.02604	0.03776	0.11298	0.16165	0.27915

Finally, Table 9 specifies the probability transducer NT by specifying the four $DT2$ models together with transition probabilities between the initial states of each of these models.

Table 9. *Probabilities specified by the NT model.*

$Pr(q_i \to q_j)$	$j = 0$	1	2	3
$i = 0$	0.9053	0.0947	0	0
1	0.1799	0.6963	0.1238	0
2	0	0.2131	0.6959	0.0910
3	0.0699	0.0413	0.1287	0.7601

	$a_1\backslash$ a_2a_3:	00	0h	01	h0	hh	h1	10	1h	11
	0	0.01577	0.01742	0.01440	0.01511	0.01215	0.01135	0.02502	0.02353	0.02786
q_0 : h		0.01478	0.01365	0.01266	0.01348	0.01324	0.01346	0.02815	0.02981	0.03442
	1	0.02701	0.02838	0.02600	0.03429	0.03158	0.03406	0.12973	0.17461	0.17809
	0	0.00962	0.01241	0.01501	0.00891	0.00753	0.01247	0.01791	0.01841	0.03530
q_1 : h		0.00818	0.00766	0.01115	0.00738	0.00952	0.01353	0.01828	0.02978	0.04405
	1	0.01946	0.01682	0.02344	0.02456	0.02668	0.03890	0.12113	0.18170	0.26020
	0	0.00406	0.00692	0.00954	0.00501	0.00372	0.00841	0.01034	0.01129	0.03430
q_2 : h		0.00391	0.00396	0.00758	0.00364	0.00707	0.01473	0.01288	0.01975	0.05058
	1	0.01250	0.01627	0.02416	0.01419	0.02071	0.04427	0.10014	0.15311	0.39698
	0	0.00302	0.00267	0.00560	0.00289	0.00249	0.00807	0.00740	0.00710	0.03195
q_3 : h		0.00297	0.00261	0.00355	0.00299	0.00271	0.00935	0.00924	0.01148	0.04296
	1	0.01035	0.01125	0.02204	0.00930	0.01289	0.04235	0.05304	0.08163	0.59810

Gregory Kucherov
INRIA/LORIA
615, rue du Jardin Botanique
B.P. 101, 54602 Villers-lès-Nancy, France
Email: Gregory.Kucherov@loria.fr

Laurent Noé
UHP/LORIA
615, rue du Jardin Botanique
B.P. 101, 54602 Villers-lès-Nancy, France
Email: Laurent.Noe@loria.fr

Mikhail Roytberg
Institute of Mathematical Problems in Biology
Pushchino, Moscow Region, Russia
Email: roytberg@impb.psn.ru

www.ingramcontent.com/pod-product-compliance
Ingram Content Group UK Ltd.
Pitfield, Milton Keynes, MK11 3LW, UK
UKHW021321180426
11947UKWH00015B/1372